Improving Concrete Quality

Improving Concrete Quality

Karthikeyan H. Obla

NRMCA
NATIONAL READY MIXED CONCRETE ASSOCIATION

CRC Press
Taylor & Francis Group
Boca Raton London New York

CRC Press is an imprint of the
Taylor & Francis Group, an **informa** business

Copublishers:

NRMCA

National Ready Mixed Concrete Association

NRMCA Publication Number: 2PE005

900 Spring Street

Silver Spring, MD

Phone: (301) 587-1400 Fax: (310) 585-4219

Email: publications@nrmca.org Website: www.nrmca.org

CRC Press
Taylor & Francis Group
6000 Broken Sound Parkway NW, Suite 300
Boca Raton, FL 33487-2742

International Standard Book Number-13: 978-1-4665-9212-4 (Hardback)

Library of Congress Cataloging-in-Publication Data

Obla, Karthikeyan H.
Improving concrete quality / author, Karthikeyan H Obla.
pages cm
Includes bibliographical references and index.
ISBN 978-1-4665-9212-4 (hardback)
1. Concrete--Quality control. I. Title.

TA440.O26 2014
620.1'36--dc23 2014001334

Visit the Taylor & Francis Web site at
http://www.taylorandfrancis.com

and the CRC Press Web site at
http://www.crcpress.com

Contents

Preface

Most concrete technologists are aware that concrete quality depends on the quality of the materials used, manufacturing, and testing. However, it is hard to find a book that is focused on the subject of *improving* concrete quality. In addition, quality measurement is not prevalent in the concrete industry, and, as a result, quality investment is not seen as generating a positive return. Over the last few years I have written a series of articles on concrete quality in the National Ready Mixed Concrete Association's Concrete InFocus magazine to address these issues. The positive comments that I received for those articles, many from strangers, encouraged me to develop this book.

The first chapter of this book discusses concrete quality measurement as well as the tangible and intangible benefits due to improved quality. Subsequent chapters discuss concrete variability due to material, manufacturing, and testing and suggest techniques to reduce them and thereby improve concrete quality. The chapter on basic statistics should provide all the background needed to understand the data analysis required for quality monitoring. Data analysis of test results and subsequent corrective action are essential for improving quality; doing a test by itself is of little value.

Improving Concrete Quality will be of significant value to the quality personnel in the concrete industry. By reading this book, they will come away with practices or tools that can be immediately employed in their operation. This book will be of substantial benefit to architects and engineers as well. As owners' representatives, quality is very important for them, and this book provides suggestions for ensuring good concrete quality. Sections of this book will interest contractors and testing laboratories as well. I hope the book creates an interest among researchers and innovators because the potential value of consistent high-quality concrete cannot be understated, and major strides still need to be made in this area. Even though this book was written with a U.S. audience in mind, the practices suggested can be easily understood and are universally applicable. Metric units are provided.

This book wouldn't have been possible without the support provided by the National Ready Mixed Concrete Association (NRMCA). Much of the reference material is based on past research conducted at the association's research laboratory. I thank the members of the NRMCA's Research Engineering and Standards Committee, which is composed of technical personnel from the concrete industry. My interactions with them over the past 10 years have helped shape my views. I thank Ken Day for the several interactions I have had with him, particularly on the CUSUM approach. I also thank my colleague Colin Lobo for reviewing my articles, and I thank Nicholas Carino for his assistance with Appendix B. I thank ACI and ASTM for allowing the use of some of the figures and my publisher Taylor & Francis/CRC Press.

The Author

Karthik Obla, PhD, PE, FACI, is vice president, Technical Services, National Ready Mixed Concrete Association (NRMCA). He has over 20 years of experience in concrete materials technology and has interests in quality control/assurance, mixture optimization, specifications, use of recycled materials, durability, and new technology. He is a fellow of the American Concrete Institute and a winner of ACI's Young Professional Achievement Award. Dr. Obla is an active member of various ACI, ASTM, and TRB technical committees and served as chair for ASTM C09.49—Pervious Concrete and is the current chair of ACI 232—Fly Ash and Natural Pozzolans. He served as a member of ACI's Concrete Research Council. He has published over 75 technical articles in journals and has presented in several international conferences. Dr. Obla earned a PhD in civil engineering from the University of Michigan, Ann Arbor, and is a licensed professional engineer in the state of Maryland. He served as vice-president and president for the ACI San Antonio Chapter. Prior to joining NRMCA he was technical manager at Boral Material Technologies.

1 How Good Is Your Quality?

Most people familiar with the concrete industry can readily identify good concrete quality practices. Regularly measuring aggregate moistures and correcting batch water is one such practice; regularly verifying accuracy of measuring devices is another. A large number of ready mixed concrete companies are making significant efforts in improving quality. However, there is always room for improvement. Most companies probably feel that they attain good concrete quality. But how does a company know for certain? For example, to evaluate a company's operational efficiencies, there are several benchmarks such as yd^3/hour (m^3/hour), delivery costs for fleets, and safety statistics that a company can measure and compare to industry averages. There are similar financial benchmarks developed by the National Ready Mixed Concrete Association (NRMCA). Are there benchmarks specific to quality of ready mixed concrete? A start can be made by the company by annually monitoring the company's cost that can be attributed to poor quality. For some companies, this may be a paradigm shift because they currently think only in terms of the cost of their quality program.

COSTS DUE TO POOR QUALITY

The cost due to poor quality may be measured by monitoring the following, typically over a 12-month period:

- Amount of rejected concrete (as a percent of the concrete produced) for noncompliance with project specifications such as slump, air content, density, and temperature, where applicable.
- Cost to repair, replace, or mitigate hardened concrete issues (cores, etc.) because concrete did not meet purchaser's or specification requirements or expectations.
- Perception of company's quality by customer through an annual customer survey (ranking as Excellent, Very Good, Good, Fair, and Poor).
- Number of quality-related claims.
- Variability in compressive strength as measured by standard deviation of the five top-selling mixtures.

The first two measures are fairly obvious. Rejected concrete, even if it is beneficially reused, is still money lost in terms of truck time and man-hours. However, returned concrete ordered in excess should not be included. Costs to repair hardened concrete can involve core tests, evaluating cracking, and so forth, and can become very expensive, even if the issues do not go to litigation. It is realized that concrete could be rejected or the producer could be asked to provide costs to address a hardened concrete issue for reasons that are not within the control of the concrete producer, such as delays in pour at the job site and job-site mixture adjustments. However, for the same contractor quality level, a better-quality producer is still likely to have a lower amount of rejected concrete and lower costs to deal with hardened concrete issues on an annual basis.

The third measure, which is the perception of a company's quality in the eyes of the customer, is also very important and should be monitored annually by concrete producers. Companies should take all possible efforts in developing and safeguarding their reputation. A customer's perception of a company's quality is formed by a number of factors, such as personnel and facilities. Typically, if customers believe that they will get a quality product with few worries, they may pay more for it. When the inevitable problems do occur, customers are likely to be more willing to look at their own operation, the testing laboratory, or other factors beyond the concrete producer. The owner, engineer, and architect are more willing to trust and even to take the advice of the concrete producer when problems do occur. They may even consult and seek the help of the concrete producer on matters of specifications and desired concrete performance. The bottom line is a happier, satisfied customer, which will inevitably lead to more business and a higher profit.

The fourth measure is fairly easy to track, and the idea behind it is that every claim involves a cost simply to address it, even if it does not lead to the actual cost that can be tracked by the first two measures.

The final measure is extremely important in any manufacturing industry, and is discussed in the following section.

WHY IS IT SO IMPORTANT TO LOWER STANDARD DEVIATION?

Variability in compressive strength, as measured by standard deviation (S), is an excellent indicator of a company's quality. The compressive strength standard deviation should be carefully monitored by every company for at least its five largest-volume mixtures or the top two selling mixtures at each plant. Table 1.1, which is a reproduction of Table 4.3 from ACI 214R-11, shows that the standards of concrete control based on general construction testing can vary from Excellent (S < 400 psi [2.8 MPa]) to Poor (S > 700 psi [4.8 MPa]). This applies when specified compressive

TABLE 1.1
Standards of Concrete Control $f'_c \leq 5000$ psi (35 MPa)

	Overall Variation				
	Standard Deviation for Different Control Standards, psi (MPa)				
Class of Operation	**Excellent**	**Very Good**	**Good**	**Fair**	**Poor**
General construction testing	<400 (<2.8)	400 to 500 (2.8 to 3.5)	500 to 600 (3.5 to 4.1)	600 to 700 (4.1 to 4.8)	>700 (>4.8)

Source: Adapted from ACI Committee 214R, "Evaluation of Strength Test Results of Concrete (ACI 214R-11)," American Concrete Institute, Farmington Hills, MI, 2011, 16 pp., Table 4.3.

strength is 5000 psi (35 MPa) or less. A low S is desirable because it will result in a lower required average strength (f'_{cr}) that a producer needs in a mix submittal for a specified strength (f'_c). Appendix A provides details about the methodology of calculating the required average strength as well as the concrete strength-acceptance requirements in accordance with ACI 318-11. The requirements are slightly different when f'_c is greater than 5000 psi (35 MPa). A lower required average strength will reduce the material costs for each class of concrete.

Row 3 of Table 1.2, indicated as f'_{cr} (ACI 318), shows the required average strength calculated for $f'_c = 4000$ psi (27.6 MPa) for different standards of concrete control. Assuming that each 200 psi (1.4 MPa) increase in f'_{cr} results in an increase in concrete materials cost of \$1/yd^3 (\$1.3/m^3), the cost savings due to the lower f'_{cr} can be estimated as shown in Row 4 of Table 1.2. It is instructive to note that improving concrete quality, that is, reducing "S" from 750 psi (5.2 MPa) to 350 psi (2.4 MPa), can result in a savings of \$3.9/yd^3 (\$5.1/m^3) in concrete materials cost due to a reduction in f'_{cr}, from 5250 psi (36.2 MPa) to 4470 psi (30.8 MPa)! In general, it can be shown that every 100 psi (0.69 MPa) reduction in S will result in material cost savings of \$1.33/yd^3 (\$1.74/m^3) when S > 505 psi (3.5 MPa), and cost savings of \$0.67/yd^3 (\$0.88/m^3) for lower S. Another way to look at this is with poor concrete quality "S" of 750 psi (5.2 MPa) as compared to 350 psi (2.4 MPa), one may have to add more cement at a cost of \$3.9/yd^3 (\$5.1/m^3) to avoid low compressive strength test results. Unfortunately, that will result in highly unoptimized mixtures and therefore is not a cost-effective practice. Poor testing quality can also increase the "S." There are ways to help improve testing quality, which is discussed in a later chapter. However, for the same level of testing quality, the producer with better quality will still have a lower "S" and is therefore still in a better position. Having a lower "S" also substantially reduces the chances of low-strength problems. Let us say a producer attained an average strength of 5250 psi for a 4000 psi specified strength level. By reducing the "S" from 750 psi to 350, the producer reduces the chances that a test result may fail the ACI 318 strength acceptance criteria from 1-in-100 to 1-in-3.3 million!

The benefit of lowering the S is even higher if one considers the traditional industry practice of reducing producer risk. If a producer exactly attained the required average strength based on the ACI 318 equations and the assumed standard deviation, there is a 1-in-100 chance that each of the two ACI 318 strength-acceptance

TABLE 1.2
Cost Savings Due to Improved Quality for $f'_c = 4000$ psi (27.6 MPa)

Quality Standards (ACI 214R)	Excellent	Very Good	Good	Fair	Poor
S, psi (MPa)	350 (2.4)	450 (3.1)	550 (3.8)	650 (4.5)	750 (5.2)
f'_{cr} (ACI 318), psi (MPa)	4470 (30.8)	4600 (31.8)	4780 (32.9)	5020 (34.5)	5250 (36.2)
Cost savings, \$/yd^3 (\$/m^3)	3.9 (5.1)	3.2 (4.2)	2.3 (3.0)	1.2 (1.5)	0.0 (0.0)
f'_{cr} (1-in-5000), psi (MPa)	4740 (32.6)	5090 (35.1)	5450 (37.5)	5980 (40)	6160 (42.4)
Cost savings (1-in-5000), \$/yd^3 (\$/m^3)	7.1 (9.3)	5.3 (7.0)	3.5 (4.6)	1.8 (2.3)	0.0 (0.0)

criteria is not met. Even though a 1-in-100 chance sounds low, what is the likelihood for at least one low test result during the course of a project involving several independent tests? ACI 214R shows that for a project with five test results (i.e., n = 5) this probability is 7.3% and it rises to 53% for n = 50! So clearly for large projects involving many tests, there is a very high likelihood for at least one low test result, providing the producer exactly attained the required average strength based on the ACI 318 equations and had no change in the assumed standard deviation. To reduce this risk, producers typically target a strength that is higher than the required average strength calculated by the ACI 318 equations. If one were to target a strength that will result in a 1-in-1000 chance of failing the ACI 318 acceptance criteria, it can be calculated that for n = 5, the probability for at least one low test result drops to just 0.75%, and for n = 50 it increases to a still acceptable 7.2%. However, for n = 250, it rises to 31.3%, which may be unacceptable. If one were to target a strength that will result in a 1-in-5000 chance of failing the ACI 318 acceptance criteria, the probability for at least one low test result drops to 7.2% for n = 250. For a 1-in-5000 chance of failing the two ACI 318 acceptance criteria, the required average strength when f'_c is 5000 psi (35 MPa) or less can be calculated as the higher of the following two equations:

$$f'_{cr} = f'_c + (2.04 \times S)$$

$$f'_{cr} = f'_c + (3.54 \times S) - 500 \text{ in psi or } f'_c + (3.54 \times S) - 3.5 \text{ in MPa}$$

The required average strength calculated by the above two equations is higher than that calculated by the ACI 318 equations and is indicated as f'_{cr} (1-in-5000), that is, Row 5 of Table 1.2. The increase in required average strength, while substantially reducing the risk of low-strength test results, clearly results in higher material cost. But it is interesting to note that reducing S leads to even more cost savings now. Reducing "S" from 750 psi (5.2 MPa) to 350 psi (2.4 MPa) can result in a savings of \$7.1/yd^3 (\$9.3/m^3) as opposed to \$3.9/yd^3 (\$5.1/m^3) in concrete materials cost.

For specified compressive strengths greater than 5000 psi (35 MPa), the standards of concrete control based on general construction testing are set on the coefficient of variation (V) instead of the standard deviation and can vary from Excellent (V < 7.0%) to Poor (V > 14.0%). This is depicted in Table 1.3, which is a reproduction of Table 4.4 from ACI 214R. Table 1.4 shows the required average strength

TABLE 1.3
Standards of Concrete Control $f'_c \geq 5000$ psi (35 MPa)

	Overall Variation				
	Coefficient of Variation for Different Control Standards, as %				
Class of Operation	**Excellent**	**Very Good**	**Good**	**Fair**	**Poor**
General construction testing	<7.0	7.0 to 9.0	9.0 to 11.0	11.0 to 14.0	>14.0

Source: Adapted from ACI Committee 214R, "Evaluation of Strength Test Results of Concrete (ACI 214R-11)," American Concrete Institute, Farmington Hills, MI, 2011, 16 pp., Table 4.4.

TABLE 1.4
Cost Savings Due to Improved Quality for f'_c = 8000 psi (55 MPa)

Quality Standards (ACI 214R) V,%	Excellent 6	Very Good 8	Good 10	Fair 12	Poor 15
f'_{cr}, psi (MPa)	8700 (60.0)	8960 (61.8)	9390 (64.7)	9990 (68.9)	11070 (76.3)
Cost savings, \$/yd³ (\$/m³)	11.8 (15.4)	10.5 (13.7)	8.4 (11.0)	5.4 (7.1)	0.0

calculated for f'_c = 8000 psi (55 MPa) for different standards of concrete control. Making the same assumption as before that each 200 psi (1.4 MPa) increase in f'_{cr} results in an increase in concrete materials cost of \$1/yd³ (\$1.3/m³), it can be estimated that improving concrete quality, that is, reducing "V" from 15% to 6%, can result in a savings of \$11.8/yd³ (\$15.4/m³) in concrete materials cost due to a reduction in f'_{cr} from 11,070 psi (76.3 MPa) to 8700 psi (60.0 MPa)! Reducing the coefficient of variation effectively results in reducing the standard deviation, thus resulting in lowering the required average compressive strengths.

IS IT WORTHWHILE NOT TO INVEST IN IMPROVED QUALITY UNDER CERTAIN CIRCUMSTANCES?

Let us say a company primarily supplies to applications where there are no job-site testing and acceptance test requirements. Then, the amount of rejected concrete related to testing will be negligible. Even though there will be no strength results to deal with, the cost to repair will be slightly higher for the producer at a lower level of quality as he will incur greater call-back costs due to aesthetics such as cracking. He is also likely to have more quality-related claims. Compressive strength is not even measured, so there is no advantage in reducing standard deviations. Companies with an excellent quality perception are likely to command a better profit.

So if the company primarily supplies to applications where there are no job-site testing and acceptance requirements, then the value of improved quality may not be so high. But even then certain basic quality functions, such as maintaining concrete yield, are bound to pay off. Further, if that company starts supplying to projects involving job-site testing and acceptance requirements, then they are likely to start seeing significant costs resulting from their lower level of quality.

2010 NRMCA QUALITY MEASUREMENT AND BENCH MARKING SURVEY

The NRMCA Quality Measurement and Bench Marking Survey, initiated in 2010 to develop quality-related metrics or benchmarks for the industry, includes four of the previously given measures as part of the survey. Selected survey results are provided as follows, and the complete survey details are available elsewhere (Obla 2011).

1. The volume of rejected concrete (as a percent of the concrete produced) for noncompliance with project specifications was reported as 1% by 45% of the respondents and less than 0.5% by 45% of the respondents.
2. The average cost to repair, replace, or mitigate hardened concrete issues because concrete did not meet purchaser's or specification requirements or expectation was reported as $0.42/yd^3 ($0.55/m^3).
3. The weighted average "S" for all company respondents was calculated as 505 psi (3.5 MPa).
4. Fifty percent of the companies reported a quality perception that was rated Excellent, Very Good, or Good. The remaining 50% of the respondents did not track their customer's perception of their quality level.
5. The average number of employees with quality/technical function, including field and laboratory technicians, was one person per 62,000 yd^3 (47,400 m^3) of production.
6. The average quality costs expended by the company, including laboratory costs and all quality staff salaries, but not including costs due to poor quality, was $1.14/yd^3 ($1.49/m^3).
7. On average, companies reported that they had one laboratory with at least one strength-testing machine for 295,000 cubic yards (225,000 m^3).
8. On average, companies reported that they conducted internal concrete strength tests about once/week.

HOW CAN A CONCRETE PRODUCER IMPROVE QUALITY?

Based on the discussion thus far, this question can be rephrased as, How do interested producers go about reducing their standard deviation? To answer this question, we have to gain a good understanding of the sources of concrete strength variation. The sources of strength variation can be broadly categorized into three parts: material, manufacturing, and testing. Under each category, there are many possible reasons for concrete variability, and some of the important ones are listed in Table 1.5. So, to reduce the strength standard deviation, the material, manufacturing, and testing variations need to be lowered. The following chapters will discuss the variation due to each category and suggest methods to reduce it.

TABLE 1.5
Sources of Concrete Strength Variation

Material	Variations in characteristics of cement, SCMs, fine aggregate (silt, grading), coarse aggregate (dust/bond), admixtures
Manufacturing	Variations in ingredient weights (water, cementitious, admixture), mixing, transporting, delivery time, temperature, workability, air content
Testing	Sampling, specimen preparation, initial/final curing of specimens, transporting, test procedures, equipment

2 Variation in Concrete Strength Due to Cement

This chapter discusses concrete strength variability due to variations of cement from a single source.

CEMENT FROM A GIVEN SOURCE VARIES BETWEEN SHIPMENTS

It is well understood that there are significant differences in strength of Type I cements from different sources. How variable is the strength of different cement shipments from the same source? One of the most exhaustive studies to address this question was conducted by Walker and Bloem (1958) at the NRMCA/NSGA (National Sand and Gravel Association) Joint Research Laboratory. Cement samples were secured from each of five sources every 2 weeks from October 1955 to October 1956 and stored in sealed containers. A sixth cement, consisting of a blend of five cement brands from the Washington area, was thoroughly mixed at the start of the program and stored in sealed containers for use as a control. Three principal series of tests were conducted at different times. The first series involved standard mortar strength tests (ASTM C109) on all samples; the second, concrete tests on selected samples; and the third, concrete and mortar tests on selected samples. In the first series, five mortar batches were made on different days with each cement sample, including the control; each round, including all sources, was made on the same day. All work was performed by the same operator. Nine cubes were molded from each batch for strength tests in triplicate at 3, 7, and 28 days of age. So, for a given cement sample at each age, 15 mortar cubes were broken (3 from each of 5 batches mixed on different days). The series involved testing of approximately 7000 2 in. (50 mm) mortar cubes. In the second series, concrete batches with a cement factor of 517 lb/yd^3 (307 kg/m^3) and mixed to a constant slump of 3 to 5 in. (75 to 100 mm) were made using samples of cement that had produced the highest, median, and lowest mortar strengths from each of the five cement sources in Series 1. For each cement sample, three concrete batches were made on different days. Concrete cylinders of size 4 × 8 in. (100 × 200 mm) were tested in triplicate at 3, 7, and 28 days. Only Series 1 and 2 are discussed below. Series 3 findings were in line with Series 1 and 2.

Figure 2.1 shows the percentage distribution of mortar strengths about the average for all three test ages. The percentage strengths at each age have been arranged in order of descending magnitude. If the different shipments from the same source were absolutely identical and the cement was essentially uniform, one should observe a perfectly horizontal line at 100%. Source 1 (control) approaches that, and the small variability is primarily due to variation attributed to testing. All the cement sources showed a greater variation than the control, with some of them better than the others. Table 2.1 summarizes some of the statistical data of the 28-day mortar strength test

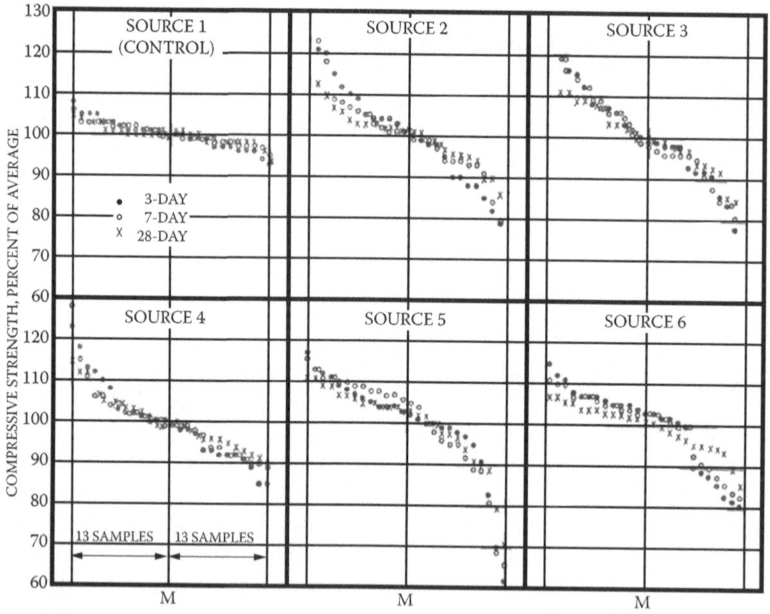

FIGURE 2.1 Percentage distribution of mortar strengths above average. (Reprinted with permission from Walker, S., and Bloem, D. L., "Variations in Portland Cement," *Proceedings, ASTM International*, West Conshohocken, PA, Vol. 58, 1958, pp. 1009–1032.)

TABLE 2.1
Statistical Summary of 28-day Compressive Strength Tests of Mortar

	Source 1 (Control)	Source 2	Source 3	Source 4	Source 5	Source 6
Average, psi (MPa)	5198 (35.8)	5355 (36.9)	4950 (34.1)	3674 (25.3)	5434 (37.5)	4582 (31.6)
s_1, psi (MPa)	133 (0.92)	132 (0.91)	125 (0.86)	79 (0.54)	137 (0.94)	111 (0.77)
s_2, psi (MPa)	119 (0.82)	131 (0.90)	127 (0.88)	81 (0.56)	137 (0.94)	114 (0.79)
s_3, psi (MPa)	123 (0.85)	339 (2.34)	362 (2.50)	244 (1.68)	520 (3.59)	245 (1.69)
v_1,%	2.6	2.5	2.5	2.1	2.5	2.4
v_2,%	2.3	2.4	2.6	2.2	2.6	2.5
v_3,%	2.4	6.3	7.3	7.3	9.6	5.3

Source: Adapted from Walker, S., and Bloem, D. L., "Variations in Portland Cement," *Proceedings, ASTM International*, West Conshohocken, PA, Vol. 58, 1958, pp. 1009–1032.

Note: Data represent 26 samples taken every 2 weeks over a year: s_1, s_2, and s_3 are, respectively, standard deviations in psi (MPa), within batch, batch-to-batch, and sample-to-sample; v_1, v_2, and v_3 are corresponding coefficients of variations.

results of the five sources. On the basis of sample-to-sample coefficients of variation (V) over the duration of the shipments, all five cement sources showed considerably greater variability than the control, as should be expected. These values of V ranged from 5.3 to 9.6% as compared with only 2.4% for the control; the values of V for the various sources were even higher with early-age strengths. The average 28-day sample-to-sample standard deviation for the five sources of cement was 342 psi. ACI 214R states that a good standard of concrete control for general construction testing is a standard deviation of 500 to 600 psi. Even though the cement-strength standard deviation may not translate directly to a concrete-strength deviation of the same magnitude, cement variation plays an important role in concrete-strength variability, and it should be clear that variation in the strength-producing property of cement from a single source cannot be ignored.

Figure 2.2 shows that C109 mortar strengths correlate very well with concrete strengths from the Series 2 work for all sources. This confirms that for a given source variation in cement strengths between shipments causes variations in concrete strengths. For the 28-day tests, the average concrete strengths of the five sources ranged from a high of 4880 psi (33.7 MPa) to a low of 4060 psi (28.0 MPa), an overall range of 820 psi (5.7 MPa). The least variable cement source resulted in a 28-day concrete-strength range of 555 psi (3.8 MPa), while the most variable cement resulted in 1205 psi (8.3 MPa). The range in 28-day concrete strength can be related to the mortar strength standard deviations as follows:

Range in 28-day concrete strength = 2.3 × mortar-strength standard deviation

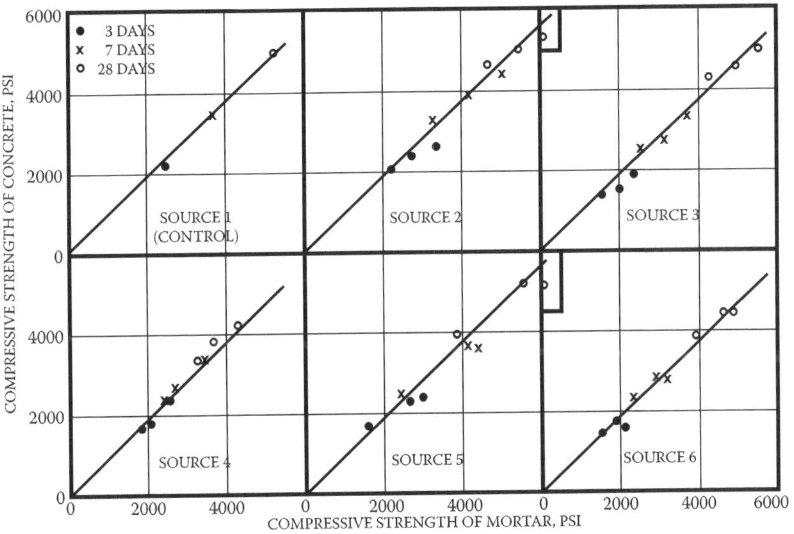

FIGURE 2.2 Relationship between compressive strength of constant-slump concrete and C109 mortar for control and the 5 cement sources. (Reprinted with permission from Walker, S., and Bloem, D. L., “Variations in Portland Cement,” *Proceedings, ASTM International*, West Conshohocken, PA, Vol. 58, 1958, pp. 1009–1032.)

This variability in concrete strengths will necessitate an increase in cement content to achieve the higher target strength to accommodate this variability. Contrary to this, Weaver et al. (1974) reported that, for the same cement source as shipments changed, there was no correlation between the C109 mortar strengths and corresponding concrete strengths. Gaynor (1993) conducted an exhaustive study of past NRMCA research and found that variations associated with testing of C109 mortar strength, as well as with concrete strength, are not insignificant. When the cement was very uniform and real cement strength variations were small, the testing variation was large enough to mask the variability that can be attributed to variation of the cement source. Additionally, when only single batches of mortar or concrete were tested, correlations between these strengths were not obvious because of testing variability. However, good correlations were obtained (Walker and Bloem 1958; Walker and Bloem 1961; Scott 1981) if several batches of mortar and concrete were prepared from the same sample, or enough consecutive samples (5 or 10) are averaged. Gaynor concluded that there was indeed an excellent correlation between C109 mortar strengths and concrete strengths made from the same cement sample.

Walker and Bloem (1958) attempted to attribute the cause for the variability in mortar strengths of the five cements' sources. Barring two instances for cement source 5, the mortar mixing water contents remained fairly constant and therefore could not explain the reasons for the variability in mortar strengths. Mortar sand grading was repeatedly checked and was found to be consistent as well. Good correlation between cement tri-calcium silicate contents and mortar compressive strengths was noted for sources 2–5. For source 6, such a correlation was not observed but the fineness was found to vary significantly. The authors concluded that variation in the tri-calcium silicate (C_3S) content was a primary factor that can be attributed to variability in mortar strengths between shipments from a given source. The authors did not notice any correlation between cement strength performance and the changing seasons.

A 1961 study by Walker and Bloem looked at 14 cements sources with two different operators and came to similar conclusions as the earlier study. In 1976 over 100 cement plants in the United States and Canada participated in a year long grab sample testing program. Cement companies collected 10 grab samples every month from shipment containers. Grab samples better represent the cement received by the concrete producer. The data from the 1976 study were statistically evaluated (Davis 1975; Klieger 1976) by a joint NRMCA/PCA (Portland Cement Association) Technical Liaison Committee and led to the development of ASTM C917, Standard Test Method for Evaluation of Cement Strength Uniformity from a Single Source, in 1980. In a later article, Gaynor (1986) analyzed the data and reported the variation in the monthly moving average strength (10 sample moving average) during the course of the year. This is reproduced in Table 2.2. A typical overdesign for concrete furnished under the ACI 318 Building Code is about 20%. Table 2.2 shows that about 40% of the cement plants experienced changes in strengths exceeding 15% during the course of the year. Half of these plants had an increase in strength. This again confirms that that variation in the strength-producing property of cement from a single source cannot be ignored.

TABLE 2.2
Changes in Cement Plant Monthly Moving Averages of Mortar Strengths

Percentage Change in 10 Samples 28-Day Strength	Percentage of Plants Exceeding 15%
5	97.1
10	86.4
15	38.8
20	17.5
25	4.9
30	1.0

Source: Adapted from Gaynor, R.D., "Cement Strength Variability and Trends in Cement Specifications in the U.S.," Eighth Congress of the European Ready Mixed Concrete Organization (ERMCO), Rome, Italy, May 1986, 8 pp.

Note: In 1976, 100 cement plants participated in year long grab sample program at 10 samples every month.

ASTM C917

ASTM C917 is the standard test method for Evaluation of Cement Strength Uniformity from a Single Source. C917 is typically done on the predominant cement sold at a cement plant. Random grab samples of cement are taken from normal delivery units, either trucks or railcars or some other point of the loading/unloading process. ASTM C183 describes grab samples as those taken in one operation. Multiple grab samples taken at prescribed time intervals may be combined to form a composite sample representative of the cement produced during that period of time. Grab samples reflect the properties of cement as shipped and will more closely explain what true variation is likely from shipment to shipment. On the other hand, composite samples may be used to develop cement mill test reports.

Typically, 10 test samples are collected in a month. All samples are tested for 7- and 28-day mortar compressive strength in accordance with ASTM C109 on standard 2-in. (50-mm) cubes. The test report includes number of samples tested, average strength, total standard deviation, standard deviation corrected for testing, and the moving average of the five most recent test results. The report will contain test data that covers a period of 3 to 12 months. At their discretion, a cement manufacturer may decide that a consistent change in the strength-producing property of cement has occurred and may start a new set of calculations; in such a case, the values calculated before, as well as after, the change should be reported.

The appendix of C917-05 has plots that summarize test data collected from 87 cement plants in the United States and in Canada in 1991. It shows that some cement plants can produce more consistent strengths than others. About 10% of the plants had a 28-day standard deviation (corrected for testing error) of more than 380 psi (2.62 MPa); 10% of plants had a 28-day standard deviation less than 180 psi (1.24 MPa); the median standard deviation of all plants tested was 280 psi (1.93 MPa). Figure 1 in ASTM C917 shows that a typical cement with a 28-day standard deviation of 310 psi (2.14 MPa), showed 28-day mortar strength varying between 5300 psi (36.6 MPa) and 6800 psi (46.9 MPa) over a period of 1 year of manufacturing. This author has reviewed a number of C917 reports and found that a 1450 psi (10 MPa) range in 28-day compressive strength is not unusual among cement plants. Some cement plants displayed 28-day compressive strength range below 1000 psi (6.9 MPa), and these plants also had standard deviations less than 225 psi (1.55 MPa).

HOW SHOULD A READY MIXED CONCRETE PRODUCER USE ASTM C917?

Ready mixed concrete producers rarely test cement, and even if they do, the results are typically obtained after the concrete containing that cement has been manufactured and utilized. Therefore, it is important for the ready mixed concrete producer to monitor the C917 reports and use them as described in the following text. This will also encourage the cement producer to pay attention to their quality practices and reduce the cement variability, which in turn will lead to lower concrete strength variability and lower materials cost.

Cement Choice

Cement-strength uniformity (ASTM C917 test data) can be considered in cement purchase decisions by the ready mixed concrete producer. Apart from ASTM C150 mill test reports, concrete producers should request to see ASTM C917 test reports over the past 12 months. Everything else being equal, a cement that has a lower standard deviation as measured by ASTM C917 will be more uniform and will generally result in a lower concrete-strength standard deviation. Based on the discussions earlier, cement with a C917 standard deviation of 380 psi (2.62 MPa) can be estimated to result in a 28-day concrete strength range of 874 psi (6.03 MPa) or ±437 psi (3.02 MPa) from the average. On the other hand, cement with a C917 standard deviation of 200 psi (1.38 MPa) can be estimated to result in a 28-day concrete strength range of 460 psi (3.17 MPa) or ±230 psi (1.59 MPa) from the average. Choosing the latter cement can result in a 200 psi (1.38 MPa) lower target average strength that can help attain a cost savings of $1/yd^3$ ($1.3/m^3$).

Better Understand Concrete Variability and Lower It

ASTM C917 test data can be used to lower overall concrete variability and hence improve quality. If, periodically (once a week), standard concrete mixtures (commonly sold mixtures at a given plant, for example) are cast and strength properties

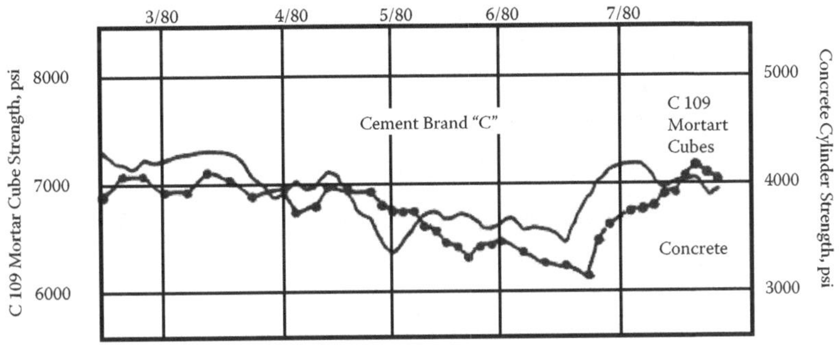

FIGURE 2.3 Running five-test average of C109 mortar cubes (C917 Report) versus running five-test average of concrete cylinders tested at 28 days. (Reprinted with permission from Scott, B. M., "Cement Strength Uniformity—A Ready-Mix Producer's View," NRMCA Publication No. 165, NRMCA, Silver Spring, MD, December 1981.)

evaluated, it becomes possible to correlate the cement-strength variation with the concrete-strength variation. Figure 2.3 shows how a concrete producer (Scott 1981) tracked the five-test moving average of C109 mortar cube strengths from the C917 test report. Superimposed is a five-test moving average of concrete cylinder test results made from a standard 450 lb/yd^3 (267 kg/m^3) cement factor, 3000 psi (20.7 MPa) concrete mixture. Both the mortar and concrete-strength curves show a downward trend from mid-March to mid-June, followed by a sharp increase. A typical 2- to 5-day time lag for rail shipments to the ready-mix plant would explain some of the apparent shift in peaks. A good understanding of the concrete-strength variation due to cement-strength variation is a first step in understanding the causes of the overall concrete-strength variation. Such an understanding can help find ways to lower the concrete variability, reduce cement factors in concrete, and attain improved quality.

Reduce Low-Strength Problems and Optimize Mixture Proportions

This can help the producer lower costs. If a process variable in cement manufacturing has changed and the cement strength is trending upward, timely communication from the cement manufacturer can help the concrete producer optimize concrete mixture proportions. On the other hand, if the cement strengths are trending downward, mixture proportions may have to be changed to prevent low-strength test results. To be effective, communication should occur as soon as 7-day mortar test results are available or even earlier if process changes have occurred at the cement plant that are known from past experience to change the strength-producing property of cement in a certain manner. In periods of high volume use, it is also possible that cement producers might switch the cement source shipped to the concrete producer.

It is important not to be alarmed by a single low or high cement 7- or 28-day strength test result. The 7-day test result is preferable as it provides a quicker opportunity to make changes as needed. A single low strength could be because of testing or genuine material variation. However, if there is a pattern of low-strength results, then it suggests a step change, that is, the average strength has reduced; and that reduction in cement strength has to be managed by changing the concrete mixture proportions to avoid low concrete strengths. The step change could be due to a testing bias but that is generally unusual and the cement manufacturer can rapidly investigate that. Most likely, the step change is due to a genuine change in the strength-producing property of the cement itself. Control charts of moving averages of 3 or 5 consecutive test results and CUSUM charts (ACI 214R; Day 2006) can help understand if a step change has occurred. Moving average control charts and CUSUM charts are discussed in more detail in the chapter on Basic Statistics. If a moving average of five consecutive test results is plotted on a control chart, statistically there is only a 2% chance that test results will fall outside $\bar{X} \pm 1.042 \times S_t$, where S_t is the total standard deviation of the cement strength and ($\bar{X}$) is the average 7-day strength, both of which are reported in a C917 test report. Example: For $\bar{X}$ = 4600 psi (31.7 MPa), S_t = 270 psi (1.86 MPa). Limits for the 2% chance can be calculated as 4600 ± 281 psi or 31.7 ± 1.94 MPa.

Troubleshoot Low-Strength Problems

When C917 data are available, concrete producers can also use it to troubleshoot low concrete strengths in evaluating whether the cause for the low concrete strength can be attributed to a reduction of the cement strength. Other factors, such as mixing water, air content, batching errors, and testing errors, should also be considered. Concrete producers should also keep 5 lb (2.25 kg) samples of cement from each shipment in sealed containers so that these can be tested if necessary. It is advisable to retain cement samples for 3 to 6 months from the date the shipment was received.

HOW SHOULD A CEMENT PRODUCER USE ASTM C917?

Cement manufacturing is an intensive process of using naturally available (variable) resources and manufacturing process variables. The cement manufacturer clearly has a goal of minimizing variability and has access to various tools and methods to monitor cement variation. A large cement producer can look at the strength variability measured according to C917 at various cement plants and try to better understand the reasons for lower variability at certain plants and duplicate them elsewhere. Sometimes, the higher strength variability attained at a plant is because of a wider variability in the raw materials of the cement itself. This was referred to by a cement producer (Floor Discussion of paper 1958) while discussing the 1958 research study. He says that some plants had a wider variation of silica content in their raw materials. If those plants attempted to control their composition of raw mixes by just the analysis for the carbonate portion, they could end up with wide variations in tri-calcium silicate contents. The cement producer can target a lower variability as measured by ASTM C917 by identifying the key strength-producing characteristics, such as tri-calcium silicate content, fineness, gypsum content, and so forth, and attempt to control them. It has been suggested that ASTM C150 itself should have limits on the extent to which some of these properties can vary (Floor Discussion 1958; Peters 1980; Mather 1975). In 1975, then-president of ASTM Bryant Mather while discussing the lower-limit-only cement strength specifications stated: "We need to do something about the variability. We need to put into the ASTM standards some way by which the user can require not just compliance with the minimum, but an assurance of some degree of uniformity." An attempt at this was to incorporate a cement strength range concept in ASTM C1157, but it was dropped because of significant confusion on how cement could be ordered.

As discussed in the previous section, as soon as the 7-day moving average of five consecutive C917 strength test results fall outside previously agreed upon control limits, the cement producer should communicate that to the concrete producer so that suitable actions can be taken. Communication can even be earlier if process changes have occurred at the cement plant that are known from past experience to change the strength-producing property of cement in a certain manner. Timely communication can reduce low-strength problems in the field and the accompanying investigating costs.

SUMMARY

1. The strength-producing property of cement from a given source varies between shipments.
2. The cement strength as measured by mortar cubes has been correlated with concrete strength. To reduce the effect of testing variability, it is important to prepare several batches of mortar and concrete from the same cement sample or compare moving averages of 5 to 10 consecutive sample test results.
3. ASTM C917 is the standard test method for the evaluation of cement strength uniformity from a single source. C917 is typically done on the predominant cement sold at a cement plant.
4. Concrete producers should monitor ASTM C917 test reports and use them to make purchasing decisions, attain lower variability concrete, and prevent and/or troubleshoot low-strength problems.
5. Cement producers should use C917 test reports to compare cement plants, better understand reasons for the strength variability, and attempt to lower it. If there are process changes at the cement plant or if the 7-day C917 strength test results fall outside agreed upon control limits, the cement producers should communicate that to the concrete producers so that they can take effective action.
6. It is important that the concrete producer and the cement manufacturer work as a team as it is in the best interest of both that concrete of good quality with low variability and reduced low-strength problems is made and placed. A good understanding of cement strength variations through effective use of ASTM C917 is essential in this regard.

3 Variation in Concrete Strength Due to Water and Air Content Variation

This chapter discusses concrete strength variability due to variations in mixing water and air contents.

MIXING-WATER CONTENT VARIATION AND ITS EFFECT ON COMPRESSIVE STRENGTH VARIATION

Compared to other ingredients, there are several sources (and potential errors) by which mixing water is incorporated in a concrete mixture. These sources and ways to ensure that the batchwater is within tolerance of the designed amount are discussed in a later chapter. ASTM C94 and ACI 117 state that the total mixing water should be within ±3% of the quantity required by the mixture proportion. ASTM C94 also requires that the added water should be within ±1% of the design total mixing water. These standards do not establish accuracy requirements for individual sources of water, such as for water added from the water tank on the concrete truck. The NRMCA Plant and Truck Certification inspection process has accuracy requirements for water measurement in the plant and for water tanks that are based on target quantities.

For the purpose of this discussion, it is assumed that for plants and construction practices operating under a good standard of control, the mixing water content can vary within ±5% of the required quantity 95% of the time. In other words for a concrete mixture with a design mixing water content of 250 lb/yd^3 (148 kg/m^3), the actual water content can vary by ±12.5 lb/yd^3 (7.4 kg/m^3). During a field evaluation conducted by the NRMCA, the mixing water content calculated from the batch records and other reported sources of water was found to range from 227 to 256 lb/yd^3 (135 to 152 kg/m^3), with an average of 243 lb/yd^3 (144 kg/m^3). This was determined from 12 loads of the same mixture. From this evaluation, 94% of the values were within the range of ±12.5 lb/yd^3 (7.4 kg/m^3). One could also consider the ASTM C94 and ACI 117 allowable slump tolerance of ±1.5 in. (38 mm) for concrete of slump greater than 4 in. (102 mm), with all else being equal, as roughly resulting from a ±5% variation in water content. As part of a different study, the author received project test data from several concrete producers. Each project test data consisted of slump, air content, temperature, and 7- and 28-day strengths measured on the same concrete sample over a period of several months. Some of that data are summarized in Tables 3.1–3.3. Table 3.1 shows that barring Producer C, the remaining four data sets met the slump requirement within tolerance exceeding 95% and

TABLE 3.1
Statistical Summary of Project Slump Test Results Sent by Five Concrete Producers

Producer ID	Percent of Measurements	Average Slump, in. and mm	Slump Range, in. and mm	Percent in Compliance with C94 Tolerance of ±1.5 in. (40 mm)
A	305	5.2 (132)	1.75–6.75 (45–170)	98.0
B	57	6.9 (175)	5.00–8.00 (125–205)	94.7
C	41	7.6 (193)	3.50–10.50 (90–265)	73.2
D1	304	4.8 (122)	2.00–8.00 (50–205)	97.4
D2	62	8.4 (213)	6.25–9.50 (160–240)	95.2

averaged 96.3%. So, the assumption that plants and construction practices operating under a good standard of control have a mixing water content variation of ±5% of the required quantity 95% of the time seems reasonable.

For a concrete mixture at a given w/cm, a ±5% variation in mixing water will have the same effect on the w/cm as a ±5% variation in the cement content. A commonly used rule of thumb is that 1 lb of cementitious material equates to a compressive strength between 8 and 12 psi (or 1 kg of cementitious material equates to a compressive strength between 0.12 to 0.18 MPa). This is a simplistic assumption that is valid only for w/cm around 0.50 to 0.60 or for concrete with 28-day compressive strength ranging from 4000 to 5000 psi (27.6 to 34.5 MPa). For lower w/cm concrete, a pound or kilogram of cement contributes lesser to compressive strength as compared to higher w/cm. (See Appendix B for more discussions.) This means that a ±5% variation in cement content will result in a ±5% compressive strength variation. Based on the discussions in Appendix B, this is valid only for w/cm around 0.50. If the mixing water content varies by ±5% of the target value 95% of the time, the effect is that the resulting compressive strength variation should be within ±5% of the average strength 95% of the time.

AIR CONTENT VARIATION AND ITS EFFECT ON STRENGTH VARIATION

ASTM C94 and ACI 117 state that for air-entrained concrete, the tolerance is ±1.5%. Table 3.2 has only two data sets with air-entrained concrete. Both the data sets show that there is a higher compliance with air-content tolerances compared to slump tolerances. Based on that data, one could conservatively assume that for plants and

TABLE 3.2
Statistical Summary of Project Air Content Test Results Sent by Five Concrete Producers

Producer ID	Number of Measurements	Average Air Content, %	Air Content Range (%)	Percent in Compliance with C94 Tolerance of ±1.5%
A	302	5.6	3.4–6.9	99.7
B	57	5.4	3.9–7.1	96.5
C	41	2.9	2.2–4.0	90.2[a]
D1	302	2.0	1.0–4.5	92.1[a]
D2	17	1.0	0.8–1.4	100[a]

[a] Nonair-entrained concrete does not have C94 air content tolerances. So ±0.75% was assumed.

construction practices, operating under a good standard of control, the air contents of air-entrained concrete can vary within ±1.5% of the average 95% of the time. Table 3.2 has three data sets with nonair-entrained concrete. ASTM C94 and ACI 117 do not state tolerances for nonair-entrained concrete. The air content variations in nonair-entrained concrete are expected to be lower than in air-entrained concrete. Assuming a tolerance of ±0.75%, the three data sets show that for plants and construction practices operating under a good standard of control, the air contents of nonair-entrained concrete can vary within ±0.75% of the average 95% of the time.

A common rule of thumb is every 1% increase in air content will reduce the concrete compressive strength by 5%. So the air content variation of ±1.5% will result in a ±7.5% compressive strength variation. In the case of air-entrained concrete, since the air content is within ±1.5% of the average value 95% of the time, the resulting compressive strength variation should be within ±7.5% of the average strength 95% of the time.

COMBINED EFFECT OF WATER AND AIR CONTENT VARIATION ON STRENGTH VARIATION

It is assumed that mixing water content, air content, and the resulting compressive strength are normally distributed.

So far we know the following:

95% of the time the water content is between 0.95W and 1.05W, where W is the target mixing-water content. Assuming a normal probability distribution, this would result in

97.5% of the time the water content is less than 1.05W and as a result,

97.5% of the time, the strength is greater than $0.95\overline{X}$, where $\overline{X}$ is the average compressive strength.

In the same way, it can be concluded that

97.5% of the time, the air content is less than (A+1.5), where A is the target air content and as a result,
97.5% of the time, the strength is greater than $0.925\overline{X}$ (1 – 0.075).

Combining the water and air content probabilities we can state that

95% (0.975 × 0.975) of the time the water content is less than 1.05 W, and the air content is lesser than (A+1.5) and as a result,
95% of the time, the strength is greater than $0.875\overline{X}$ (1 – 0.05 – 0.075)

This is somewhat of a conservative approach since this approach assumes that high slump and high air content are mutually exclusive events, whereas in reality, a high water content is likely to lead to a high air content and vice versa.

Let us assume the compressive strength test results are normally distributed with a standard deviation (S). So, for the 95% probability it follows that:

$$\overline{X} - 1.64S = 0.875\overline{X}.$$

Simplifying, we get

$$S = 0.0762\overline{X}$$

If the average measured compressive strength ($\overline{X}$) is 5000 psi (34.5 MPa), then S = 381 psi (2.63 MPa). ACI 214R states that S < 400 psi (2.76 MPa) would denote an excellent standard of concrete control for general construction testing. However, it is important to realize that the value of S calculated here (albeit, conservatively) is solely due to variation in the water and air contents. Variation due to cement, supplementary cementitious materials (SCMs), aggregates, admixtures, mixing, transporting, delivery time, temperature, curing, and testing is likely to significantly increase this value.

DISCUSSION

The above analysis can be used to illustrate why nonair-entrained concrete has a lower variability in compressive strength. It was shown that nonair-entrained concrete typically has an air content variation of ±0.75%, which can result in a ±3.75% variation in compressive strength. When this is added to the water content variation as in the above example, it results in S = 267 psi (1.84 MPa) for the average measured compressive strength of 5000 psi (34.5 MPa). However, there are instances where in the combinations of some admixtures and cementitious materials (even nonair-entrained concrete mixtures) may attain high air-content variations, thus leading to potentially higher strength variations. For example, Table 3.2 shows that for producers C and D1, the air contents of their nonair-entrained concrete mixtures exceeded ±0.75% of the average about 10% of the time.

TABLE 3.3
Statistical Summary of Project Test Results Sent by Five Different Concrete Producers

Producer ID	Coefficient of Variation, %				S of 7 Day Strength, psi (MPa)	28-Day Strength		
	Slump	Air Content	Temperature	Density		Average, psi (MPa)	S, psi (MPa)	Predicted S, psi[a] (MPa)
A	12.9	8.2	9.2	NA	485 (3.35)	5578 (38.5)	438 (3.02)	307 (2.12)
B	10.6	15.0	9.9	0.94	624 (4.30)	4952 (34.2)	536 (3.70)	362 (2.50)
C	22.2	15.9	17.1	0.84	1701 (11.73)	9450 (65.2)	1258 (8.68)	890 (6.14)
D1	12.3	22.5	8.9	1.03	594 (4.10)	5213 (36.0)	474 (3.27)	281 (1.94)
D2	9.0	18.0	7.3	0.36	1167 (8.05)	6632 (45.7)	938 (6.47)	294 (2.03)

[a] Predicted S based on slump and air content compliance probabilities.

Table 3.3 is a statistical summary of project test results sent by the five different concrete producers. The coefficient of variation of slump, air content, temperature, density, and standard deviation of 7- and 28-day strengths are provided. The following observations can be made:

1. The standard deviation of 7-day strength is higher and correlates well with the standard deviation of the 28-day strength.
2. The standard deviation of the 28-day strength varies from 438 psi (3.02 MPa) to 1258 psi (8.68 MPa).
3. The variability of slump and air contents correlates with the measured variability of the 28-day strength. The predicted 28-day strength standard deviation is calculated from the percent in compliance with slump and air-content tolerances, as discussed in the previous section. The slump variation is converted to a water content variation. The predicted 28-day strength standard deviations suggest that producers A and D1 have the lowest variability, followed by producer B and finally producer C. Producer D2 does have a low predicted S but reported a high measured 28-day strength S. Producer D2 had combined the test data from several different laboratories, and the resulting testing variations have made it harder to interpret that set of data. Producer C had the lowest percent in compliance with C94 slump tolerance limits, which resulted in the highest coefficient of variation of concrete slump. Producer C also had the highest coefficient of variation of concrete temperature. This suggests that producer C might have a high variation in water content, which can explain the high measured S of

28-day strength. Producer A, on the other hand, attained the highest percent in compliance with slump and air content and ended with the lowest measured S of 28-day compressive strength.

SUMMARY

Variation in mixing-water content and air content has been shown to contribute significantly to the variation in compressive strength. So in order to attain a low standard deviation of 28-day strength, it is important for the producer to target a low mixing-water and air content variation.

4 Mixing-Water Control

This chapter discusses how a producer can manage the various factors that will reduce the variability of mixing-water content in concrete mixtures.

SOURCES OF WATER

Compared to other ingredients, there are several sources (with potential errors) by which mixing water is incorporated in a concrete mixture. ASTM C94 and ACI 117 state that the total mixing water should be within ±3% of the quantity required by the mix design. This mixing water includes water added to the batch, ice added to the batch, water occurring as surface moisture on the aggregates, water introduced in the form of admixtures, and any washwater retained in the truck drum. In addition, mixing water may also be added at the slump rack and at the job site. The primary sources of mixing-water variations and methods by which they can be controlled to be within a tight tolerance are discussed next.

WASHWATER IN TRUCK MIXER DRUM FROM PREVIOUS LOAD

When concrete is discharged at the job site, a mortar coating adheres to the surface of the mixer drum. Depending on the situation, type of mixture or company practice, the concrete mixer truck operator may wash the interior of the drum at the job site or on arrival at the plant. In some cases, this water may not be discharged, and it may remain in the truck when the next load is batched. Washing the mixer drums between every load leads to several possible problems:

1. It consumes a lot of water.
2. If discharged, it generates process water at the concrete plant that will have to be managed in compliance with environmental regulations.
3. If not discharged, water (anywhere from 5 to 30 gallons or 18.9 to 113.4 liters) may remain in the mixer drum when the next load is batched.

ASTM C94 recognizes the last possibility and states a condition that the washwater retained in the drum for use in the next load should be accurately measured. If that is not possible for compliance with ASTM C94 truck operators should be educated on the importance of disposing of all of the washwater prior to batching a fresh concrete load. The process of discharging the mixer washwater into a wash-out pit may take time, but it is an important step toward achieving quality concrete. The more common industry practice is to not wash out the mixer drum after every load. The mortar coating will not significantly change the water content or any of the properties of the new load of concrete. Washing the inside of a mixer drum may be necessary in hot weather conditions when it can lead to undesirable hardened concrete buildup inside the drum, when using special concrete-containing color or fibers, or when required by company policy.

Batchwater

Batchwater is defined as the water added to the ready mixed concrete truck. Generally, this constitutes the primary source of mixing water. Water can be measured by volume using water meters or volumetric tanks or by mass in scales. ASTM C94 requires that the added water should be accurate to within ±1% of the design total mixing water. Water is most commonly measured by water meters, which are volumetric batching devices for water. In some plants, water may also be measured by mass through scales. The NRMCA plant certification requires that volumetric water measuring devices should be accurate to within ±1.5% of the desired amount of water, which corresponds approximately to an accuracy of ±1% based on total mixing water for typical aggregate moisture levels. Scale accuracy requirements are tighter. The NRMCA measuring tolerance is more easily verified during a batching process or by reviewing batch records. The NRMCA plant certification process requires that measuring devices be verified for accuracy not less than once every 6 months or when there is reason to assume their accuracy is in question. It is one thing to have a well-calibrated water-batching device. It is additionally important to ensure that the added water in every concrete load that is shipped from the plant is within target range. Batching errors do occur from time to time. All computerized batch panels have some form of error monitoring and alerting system, which may be overridden and accepted with a keystroke. Automated evaluation and alerting systems available nowadays can send email alerts to Quality Control/Quality Assurance (QC/QA) personnel when a load is found to be outside prescribed or company-set target ranges. Improving batching accuracy will be discussed at great length in a later chapter.

Most chemical admixtures added to concrete contain in excess of 60% water. It is important to consider the water content contributed by chemical admixtures, particularly those used in high dosages such as corrosion inhibitors or accelerators. In general, if the admixture dosage is around 1 gallon/yd^3 (5 liter/m^3) or greater, batch water compensation should be made to account for water incorporated with admixtures. For the admixtures used in high dosages, if the dosages change significantly between loads and the water from the admixtures had not been considered, then it may lead to high mixing-water variations.

Free Water from Aggregates

Free water from aggregates constitutes a portion of the mixing water. If the coarse aggregate and the fine aggregate have a free moisture content of 1% and 5%, respectively, assuming a typical concrete mixture that contains 1800 lb/yd^3 (1068 kg/m^3) of coarse aggregate and 1200 lb/yd^3 (712 kg/m^3) of fine aggregate, the contribution that the aggregates provide to the mixing water is 78 lb/yd^3 (46 kg/m^3). If the total mixing water is 270 lb/yd^3 (160 kg/m^3), the free water from the aggregates contributes 29% of the mixing water in this example. Therefore, even a 1% error in estimating the aggregate moisture content can cause a 12–18 lb/yd^3 (7.1–10.7 kg/m^3) variation in the mixing-water content, which can be roughly equal to a 300 psi (2.1 MPa) change in 28-day compressive strength. To address this, many concrete producers use moisture probes in the fine aggregate bins. These moisture probes provide real-time moisture measurements of the fine aggregate, based on which the added water is adjusted automatically or manually. The NRMCA plant certification requires that moisture probes be verified for accuracy not less than once every 6 months or when there is an assumption of error. Alternatively, aggregate moisture should be determined at the concrete plant using a hot plate or a microwave oven, according to ASTM C566. If such a process is followed, the NRMCA plant inspector guide suggests that the aggregate moisture be measured not less than three times per week. Aggregate moistures can vary, depending on aggregate storage methods or frequency of receiving shipments. Moisture probes are not commonly used in coarse aggregate bins due to wear. Coarse aggregate moistures should be periodically measured using a microwave oven or a hot plate. Since they constitute the largest quantity of ingredients batched, a small change in the coarse aggregate moisture can result in a large change in the mixing-water content. Another important factor is the procedures used by the loader operator to feed aggregates from stockpiles to the batch plant. The key is to avoid loading aggregates with variable moisture and to ensure that the aggregate being batched is representative of the moisture content assumed in the batching process. In general, working the complete face of a stockpile tends to average out the moisture content and minimize variations. Finally, if aggregate batch weights are substantially out of target range (10% or more), they can also lead to variations in mixing-water content. Generally, such grossly out-of-tolerance aggregate batch weights can easily be identified and rectified.

Water Added at Slump Rack

Almost all concrete is accepted at the job site based on a certain slump level or visual consistency to facilitate ease of placement, regardless of whether the project specification includes a slump requirement. After loading all of the materials, truck mixer operators move the truck to the wash-down area or slump rack to clean the loading hopper, chute, and sides of the mixers before leaving the plant. At this point, they may visually estimate the slump and might retemper the load to achieve the desired mixture consistency, including some anticipated slump loss during transit. In some cases, actual slump measurements may be made and/or slump meters on truck mixers may be used. But typically this adjustment is done visually as determined by the truck operator. Water added at the slump rack can be a large source of variation in mixing water in concrete and needs to be managed by proper company policy and education of truck mixer operators. Some options to control this include using timers on water hoses; using high-pressure, low-volume spray bars; installing water meters on the water lines; or using water from the truck water tank for retempering to obtain a measurement of the added water. The goal should be to minimize the water added to the mixture at this point and to be able to record the quantity of water added to the mixture. It is important to educate the truck mixer operators to minimize the water entering the mixer drum and to record the amount of water added.

Automated water-addition devices installed in truck mixers are also available. These devices introduce recordable quantities of water to the concrete mixture based on calibrated slump meters that measure the mixture consistency. These devices can be controlled to shut off when the mixing water reaches the quantity permitted for the mixture. Automated admixture-addition devices are also being explored.

Water Added at Job Site

When the truck arrives at the job site and the concrete is about to be unloaded, it is quite possible the concrete slump is lower than that specified or desired by the contractor for placement. This depends on the time elapsed from batching, ambient and concrete temperature, and mixture proportions used, among other factors. ASTM C94 allows a onetime addition of water so that concrete slump can be brought to specified levels, provided the design mixing-water content is not exceeded. A onetime addition of water may be several distinct additions of water, and all water addition should be completed within 15 minutes. Since 2013, ASTM C94 has also started to allow the use of automated water-addition devices in truck mixers as well. It is common practice for producers to hold back water to permit job-site water addition (NRMCA CIP 26). This water addition should be done before any significant quantity of concrete is discharged, and C94 allows for slump and air tests on a preliminary sample to facilitate this. C94 requires that the water added by the receiver of concrete and his initials be recorded on the delivery ticket. At the request of the purchaser, C94 requires that the delivery ticket has all the information required for calculating the total mixing water. It is a good practice for producers to note the amount of water that can be added at the job site so that design mixing-water content is not exceeded. ACI and ASTM standards do not establish accuracy requirements for water added from truck water tanks. The NRMCA Plant Certification has accuracy requirements for measurement devices on truck water tanks. Sight gauges have to be accurate within ±1 gallon (±3.78 liter), and water meters have to be accurate within ±2% of target water.

VARIATIONS IN MIXING-WATER DEMAND

As mentioned earlier, concrete at the job site is expected to be within a specific slump range before it can be placed. Temperature, delivery time, air content, and a host of other factors affect the mixing-water demand of a given concrete mixture that needs to meet a specific slump level.

Table 4.1 summarizes the effect of temperature and delivery time on mixing water demand from a previous NRMCA research study (Gaynor et al. 1985). The study shows that, if a mixture had been designed at a concrete temperature of 65°F (18.3°C) and a delivery time for 20 minutes, an increase in temperature to 95°F (35.0°C) and delivery time to 90 min can cause an increase in average mixing-water content of 33 lb/yd^3 (43 kg/m^3) to maintain the same slump. This variation in mixing water can

TABLE 4.1
Average Mixing-Water Increase Due to Temperature and Delivery Time

Condition	Average Mixing-Water Increase,[a] lb/yd³ (kg/m³)
Delivery time maintained at 20 min and concrete temperature increased from 65°F (18.3°C) to 95°F (35°C)	12 (7.1)
Delivery time maintained at 90 min and concrete temperature increased from 65°F (18.3°C) to 95°F (35°C)	19 (11.3)
Temperature maintained at 65°F (18.3°C) and delivery time increased from 20 to 90 min	14 (8.3)
Temperature maintained at 95°F (35°C) and delivery time increased from 20 to 90 min	21 (12.5)
Temperature increased from 65°F (18.3°C) to 95°F (35°C) and delivery time increased from 20 to 90 min	33 (19.6)

Source: Adapted from Gaynor, R.D., Meininger, R.C., and Khan, T.S., *Effect of Temperature and Delivery Time on Concrete Proportions*, Temperature Effects on Concrete, ASTM STP 858, T.R. Naik, Ed., American Society of Testing Materials, Philadelphia, June 1985, pp. 68–87, www.astm.org.

[a] Mixing-water increase was averaged over eight different mixture types (two different cements, two different strength levels, with and without fly ash). In all cases, mixing-water content was adjusted to attain a slump level of 4 ± 1 in. (102 ± 25 mm) at discharge.

clearly lead to increased standard deviation in strengths and low-strength problems. The concrete producer has limited control over concrete temperature and very little control over the delivery time.

The producer can reduce the mixing-water variations due to temperature variations through the following steps:

1. Table 4.1 suggests that, for every 10°F (5.6°C) increase in concrete temperature, the mixing-water demand increases by about 5 lbs/yd^3 (3 kg/m^3), which is about 2% of the mixing-water amount used, typically assuming a delivery time of 60 min. So the producer could increase the dosage of water-reducing admixtures as the concrete temperature increases by more than 10°F (5.6°C).
2. Concrete mixtures designed at a concrete temperature of 65°F (18.3°C) cannot be used when the concrete or ambient temperature exceeds 85°F (29.4°C). When the temperature reaches and exceeds 85°F (29.4°C), adjustments, such as the use of retarders and increased amounts of pozzolans like Class F fly ash or slag cement, allow for longer slump retention without excessive water addition.

Variations in mixing-water demand due to delivery-time variations are much harder to control since the producer cannot predict the delivery time of any particular load in advance. The challenge is that the travel time may be unpredictable due to traffic in urban environments. Even if travel time is consistent, the time at which the concrete gets poured will vary, depending on the contractor's work schedule. Based on Table 4.1 results, concrete producers should aim to target a slump of 2–3 in. (51–76 mm) higher than the minimum permitted at the job site or hold back about 2–3 gal/yd^3 (7.6 to 11.3 L/m^3) of mixing water to compensate for slump loss that can occur due to the delivery time. Even though that will help attain consistent slumps at the job site, the mixing-water content will still vary due to delivery-time variations. There will be no variations in mixing-water demand due to delivery-time variations if the concrete mixtures do not exhibit any slump loss. As explained earlier, use of retarders and increased amounts of pozzolans like Class F fly ash or slag cement allows for longer slump retention, but one has to be careful adopting those techniques when concrete and ambient temperatures are below 85°F (29.4°C) due to potentially delayed setting times. Job-site admixture (water reducer) addition may also help reduce the variations in mixing-water demand due to delivery-time variations, provided qualified personnel are available to administer it. Automated admixture (water reducer) addition devices can also be considered when available.

ASTM C94 and ACI 117 state that for air-entrained concrete, the tolerance is ±1.5%. Figure 4.1 extracted from the NRMCA Concrete Technologist Certification Instruction Manual shows the typical reductions in mixing-water content expected due to the entrained air content. Added air illustrated in the figures is the difference between the target air content in air-entrained concrete and that present as entrapped air in nonair-entrained concrete with the same materials. Air entrainment causes greater water reductions for low-cement factor or higher w/cm concrete mixtures. If two successive concrete loads have a total air content at the concrete plant of 7.5% and 4.5% (just meeting the C94 tolerance of 6 ± 1.5%), the added air for each load

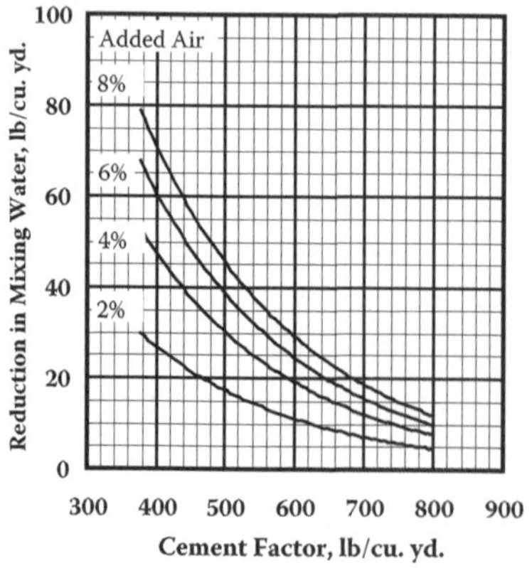

(a)

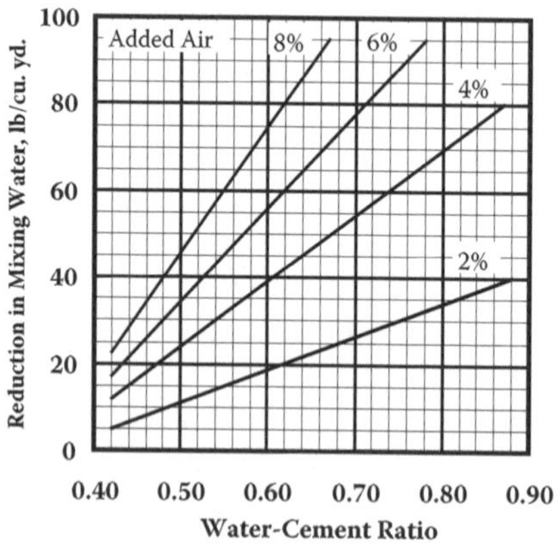

(b)

FIGURE 4.1 Effect of added air on mixing-water demand. (Reprinted with permission from NRMCA Concrete Technologist Certification Instruction Manual, "Outline and Tables for Proportioning Normal Weight Concrete—Advanced Version," NRMCA, Silver Spring, 2010.). (a) When cement factor is known; (b) when w/cm is known.

can be calculated as 6.0% and 3.0%, and therefore the difference in mixing-water demand can be estimated from Figure 4.1b as 20 lbs/yd^3 (11.9 kg/m^3), assuming a design w/cm of 0.50 for that concrete mixture. There is a silver lining in this, though. Generally, a higher air content results in a lower mixing-water demand, so the resulting strength reduction due to the higher air content is somewhat reduced. Once again, job-site admixture (water reducer) addition or the use of automated admixture (water reducer) addition devices can help reduce the variations in mixing-water demand due to air content variations.

In addition to temperature, air content, and delivery time, other variables such as variations in cementitious shipments, presence of fines in aggregate, and dust on the aggregate surface can also influence the mixing-water demand.

EFFECT OF MIXING-WATER CONTENT, MIXING-WATER DEMAND ON MEASURED SLUMP

For a given concrete mixture, if 10 different truckloads batched have slump varying over a tight range (within 1 in. or 25 mm, for example), would it be reasonable to expect that the mixing-water content is controlled to a tight range? The answer is yes, provided the mixing-water demand stays the same.

Consider the job site data for the following two truckloads that used identical materials and mixture proportions:

Load A. Concrete temperature = 65°F (18.3°C), air content = 7.5%, delivery time = 20 minutes, slump = 4 in. (102 mm).
Load B. Concrete temperature = 65°F (18.3°C), air content = 4.5%, delivery time = 90 minutes, slump = 4 in. (102 mm).

From the discussions in the previous section, it can be quickly estimated that Load B is likely to have about 35 lb/yd^3 (20.8 kg/m^3) higher mixing-water demand than Load A, but since the slumps measured at the job site are identical, the actual mixing-water content of Load B should have been about 35 lb/yd^3 higher (20.8 kg/m^3)! Using the same argument, if the temperature of Load B is 20°F (11.7°C) higher than that of Load A, it can be estimated that the actual mixing-water content of Load B should have been nearly 50 lb/yd^3 (29.7 kg/m^3) higher! In addition, other variables such as variations in cementitious shipments, presence of fines in aggregate, and dust on the aggregate surface can also influence the mixing-water demand, thus resulting in loads that are similar in slump but that can differ considerably in mixing-water content. With this kind of potential difference in mixing-water content for two loads with the same slump, one wonders about the basis of rejecting concrete at the job site on the basis of slump.

The above example clearly shows that the measured slump is due to a combination of the mixing-water content in the concrete batch as well as the mixing-water demand of that concrete batch. The effect of the concrete batch's mixing-water content and mixing-water demand on the measured slump can be summarized as shown in Table 4.2. While it is easy to understand how an increase in mixing-water content leads to a higher measured slump, it is interesting to note that an increase

TABLE 4.2
Effect of Mixing Water, Mixing-Water Demand on Measured Concrete Slump

Mixing Water in the Batch	Mixing-Water Demand of the Batch	Effect on Measured Concrete Slump of the Batch
Higher than target range	↔	↑
	↓	↑↑
	↑	↔
	↑↑	↓
Lower than target range	↔	↓
	↑	↓↓
	↓	↔
	↓↓	↑
Within target range	↔	↔
	↑	↓
	↓	↑

in mixing-water content may also lead to the same, and even a lower, slump, provided the mixing-water demand increases. Similarly, a decrease in mixing-water content may also lead to the same, and even higher, slump, provided the mixing-water demand decreases.

PLANT TESTS FOR QUALITY ASSURANCE

Slump measurement at the plant (not at the job site) prior to any water addition at the slump rack can serve as a plant quality assurance tool. To begin with, the mixing-water variations are likely to be lower since the water added at the slump rack and at the job site is not considered; only the other three potential sources of water, such as washwater in the truck drum, free moisture from aggregates, and batchwater, are considered. Also the variation of the mixing-water demand is also likely to be lower since the delivery time is not considered; only the other sources of mixing-water demand variations such as temperature, air content, characteristics of the materials, and batched quantities, are considered. So slump variations at the plant can be tracked to better understand the variations in mixing-water content.

Let us say a producer targets a slump range at the plant. If the slump measured at the concrete plant prior to any water addition at the slump rack is out of the target slump range for that mixture from Table 4.2, it is clear that either the mixing-water content is not within target range or the mixing-water demand for that mixture has changed. In Table 4.2, out of 11 possible cases, there are 2 cases where the slump may not be out of the target range, yet the mixing-water content is not within target range. These 2 special cases will be discussed later.

When a slump measurement is out of the target range, the following systematic investigation can help explain the cause of that problem:

1. Was the batchwater out of the target range? (Typically, batchwater is 200 lb/yd^3 [119 kg/m^3], so the maximum possible variation in mixing water is ±3 lbs/yd^3 [1.8 kg/m^3] if batchwater is within target range.)
2. Was the correct aggregate moisture content used in calculating batch weights? (It is a good idea to conduct a quick aggregate moisture test either through the microwave oven or a hot plate or using a Speedy moisture meter [ASTM C70]).
3. Was there washwater in the mixer drum?
4. Was the air content out of tolerance? (Typically, the air content tolerance is ±1.5%; this can lead to a variations in mixing-water demand of ±10 lbs/yd^3 [5.9 kg/m^3].)
5. Was the concrete temperature more than 20°F (11.7°C) different from the design concrete temperature for that mixture? (From Table 4.1, it can be estimated that, for every 10°F [5.9°C] increase in concrete temperature, the mixing-water demand should increase by about 4 lbs/yd^3 [2.4 kg/m^3], provided the slump is measured at the plant within 20 minutes after batching the materials.)
6. Was the batched quantity of any of the solid materials varying by more than 10% of the target values? Was the water-reducing admixture dosage out of tolerance by more than ± 20%?

If the answer to these questions is no, then changes in material characteristics, such as variations in cementitious shipments, excessive fines in aggregate, and dust on the aggregate surface, should be evaluated. By tracking slump variations on a daily basis and investigating out-of-tolerance results, the underlying cause for the slump variations can be understood. Concrete producers who already follow good quality practices have an easier job answering these questions and understanding the root cause of the slump variations. If the root cause is mixing-water variations, then an attempt should be made to bring the mixing-water content within the target range. If the root cause is mixing-water demand variations, an attempt should be made to bring the slump within the target range without changing the mixing-water content. This can be done by changing the dosage of chemical admixtures.

Table 4.2 shows that in two cases even when the slump is within the target range the mixing-water content is not within the target range; in one case, a lower mixing-water content is accompanied by a lower mixing-water demand, and in another case, a higher mixing-water content is accompanied by a higher mixing-water demand.

Only the second case is a problem as it can potentially lead to low-strength problems. However, even that is not always the case. For example, if the mixing-water demand increased due to a low air content, and this was accompanied by a high mixing-water content, the resulting compressive strength may not be necessarily low. However, if the mixing-water demand increased due to changes in materials characteristics such as excessive fines in aggregate, and this was accompanied by a high mixing-water content, the resulting compressive strength may be low and a cause for concern! This specific scenario cannot be identified by merely monitoring concrete slump variations. If density tests are also conducted, a consistently lower density test result may help identify this scenario.

It is impractical to measure slump at the plant prior to water addition for every load that leaves the plant. It may be sufficient to do that every 5 to 10 loads. For the remaining loads, visual slump estimations can be made by the truck mixer operators. These slump estimations and measurements should be captured on a daily basis by QC/QA personnel, and slump test results that are outside the target slump range should be investigated as discussed above. If the mixer truck is fitted with automated slump measuring devices, it becomes easier to analyze and address mixing-water variations.

SUMMARY

The different sources of mixing water, mixing-water demand variations, and the means to control them are summarized in Table 4.3. By following the various steps, a producer focused on quality should be able to maintain good control of the mixing-water content and meet the ASTM C94 and ACI 117 tolerance of ±3% of the quantity required by the mixture proportions. On a daily basis, QA/QC personnel can utilize the slump measurements at the plant prior to water addition at the slump rack as a quality assurance tool to ensure that the mixing-water content is within target range.

TABLE 4.3
Sources of Mixing Water, Mixing-Water Demand Variations, and Means to Reduce Them

Sources of Variations	Recommended Practice
Washwater in truck mixer drum from previous load	Avoid washing the mixer drum after every load. If that is not always feasible, educate truck mixer operators to discharge all of the washwater prior to batching a fresh concrete load, or measure the washwater retained in the drum for use in the next concrete load.
Batchwater	Calibrate water measurement devices to be accurate to ±1.5% of the target quantity and verify accuracy at least every 6 months, and more often if necessary.
	Educate batchmen not to override batching error alerts.
	If possible, invest in an automated alerting system for out-of-batch tolerance that can send email alerts to QA/QC personnel to address a load before it reaches the customer.
Aggregate moisture	Have moisture probes in fine aggregate bins if possible and verify for accuracy at least every 6 months.
	Do not ignore measuring coarse aggregate moisture content.
	If probes are not used, use a hot plate or a microwave oven. Measure aggregate moisture and update batch computers at least three times a week.
	Adjust added water in all loads to account for changes in aggregate moisture content.
	Stockpile aggregates and load aggregates into the plant to minimize variations.
	All of the above are NRMCA plant certification requirements.
Water added at the slump rack	Establish company policy and equipment for slump rack water adjustment. Consider adding the water from truck water tank or install water meters in water lines. Record the amount of water added.
Water added at the job site	ASTM C94 allows a onetime addition of water to enable slump adjustments.
	Note the amount of water that can be added at the job site so that design water content is not exceeded.
	Truck operators should be required to record job-site water addition and to obtain the signature of the purchaser or representative on the delivery ticket. Ensure measuring devices on truck water tanks comply with the NRMCA Plant Certification accuracy requirements.
Temperature, delivery time, air content, and other factors	Increase the dosage of water-reducing admixtures to target a 2% mixing-water reduction for every 10°F (5.9°C) increase in the concrete temperature over the design temperature.
	Concrete mixtures designed at a concrete temperature of 65°F (18.3°C) should appropriately be modified when the concrete or ambient temperature reaches or exceeds 85°F (29.4°C).
	Consider mixtures with good slump retention performance without delayed setting times to address delivery-time variations.
	Consider job-site admixture (water reducer) addition with qualified personnel at the job site to address delivery time and air content variations.
	On a daily basis, QA/QC personnel can utilize the slump measurements at the plant prior to any water addition after batching as a quality assurance tool to ensure that the mixing-water content is within tolerance.

5 Variation in Concrete Strength and Air Content Due to Fly Ash

This chapter discusses concrete strength and air content variability due to variations of fly ash in a single source.

VARIABILITY OF FLY ASH SHIPMENTS FROM GIVEN SOURCE

A 2012 NRMCA survey (Obla et al. 2012) reported that fly ash was used in 60% of all ready mixed concrete placed in the United States. It is the most widely used SCM in the United States and constitutes 15% of all cementitious content used in ready mixed concrete. Due to sustainability and performance benefits, the use of fly ash continues to grow. Fly ash is an industrial by-product produced due to coal burning in an electric power utility. Electric utilities are primarily focused on optimizing power generation and are not concerned with fly ash quality or variability. Fly ash properties may change, depending on type and origin of coal used, blends of coal used, changes in the burner, and other factors.

Back in the late 1980s, the NSGA/NRMCA Joint Research Laboratory developed a draft Standard Practice for Evaluation of Uniformity of Fly Ash from a Single Source (NRMCA Draft Standard Practice 1988). The practice was used to conduct a detailed experimental study on four different fly ash sources—three ASTM C618 Class F fly ashes (F1, F2, F3) and one Class C (C1) fly ash. From each source, 10 fly ash samples were procured per month, each on a different day, for 6 months, spread evenly throughout the period, for a total of 60 samples. The fly ash samples were procured from ready mixed concrete plants with each sample representing a different fly ash shipment. A large stock of cement was also procured and kept in sealed containers to maintain its properties during the course of the whole experimental program. The following tests were conducted on each fly ash sample:

1. *Moisture content* loss on ignition (LOI); *fineness* as percentage passing the No. 325 sieve; *density*; *water requirement*; 7-Day Strength Activity Index (SAI)—all tests according to ASTM C311.
2. *Uniformity of color* compared to the previous shipment (NRMCA Draft Standard Practice 1988). See Appendix C for details.
3. *Foam Index Test* (Meininger 1981). See Appendix C for details.
4. *Mortar air content and loss* in air, based on a modified C311 (Lane 1991). See Appendix C for details.

Duplicate tests were run on three randomly selected fly ash samples from each group of 10. All the tests were conducted on the duplicate samples with the exception of uniformity of color. For the strength activity index, one control reference cement mortar batch was cast for every 10 mortar batches cast with the fly ash samples. The same reference cement mortar strength was used for calculating the SAI of the 10 fly ash samples. Since cement from one shipment was being used for the reference cement mortar cubes, the variation in cement mortar strengths provided an indication of the mortar testing variation. A statistical analysis of some of the experimental results is compiled in Tables 5.1 to 5.4.

AIR ENTRAINMENT

Tables 5.1 to 5.4 show that the LOI varies four to seven times between shipments for each of the four fly ash sources over the tested 6-month period. The initial mortar air content for a specific air-entraining admixture (AEA) dosage also varied between shipments for each of the sources. It varied the least for fly ash C1 (13.1% to 14.6%); it varied the most for fly ash F3 (11.1% to 18.4%) and varied moderately for fly ash F1 (8.1% to 13.4%) and fly ash F2 (9.7% to 13.5%). So in terms of its effect on initial mortar air content, the different fly ash sources can be ranked in terms of more to less variable as follows: F3>F1>F2>C1. Figure 5.1a–d shows that for fly ashes F1, F2, and F3 the loss in mortar air content with time increased as the initial mortar air content decreased, thus suggesting that when the fly ash affected the mortar air content, it is likely to lead to poorer stability of the entrained air as well. As expected, fly ash C1 had the lowest loss in air content (up to 1%), and fly ash F3 had the highest loss in air content (up to 6%) followed by F1 (up to 5%), and F2 (up to 3.5%). There were a few instances of a slight gain (<1%) in air content with time as well. For all further discussions, including tables and figures in this chapter, the term *mortar air content* will refer to the initial mortar air content.

Figures 5.2a–5.5a show the relation between LOI and mortar air content for samples of fly ashes F1, F2, F3, and C1, respectively. Figures 5.2b–5.5b shows the relation between foam index and mortar air content for samples of the same fly ashes. Fly ashes C1 and F2 did not have good correlations between LOI versus mortar air content and foam index versus mortar air content. One possible reason could be that fly ash C1, and to a.lesser extent fly ash F2, had a very low variation in mortar air content between the different samples to begin with. Fly ash F3 had the best correlations, whereas fly ash F1 had acceptable correlations for LOI and mortar air content. Fly ash F3 had the largest change in LOI content. It appears that both LOI and foam index tests can be useful at estimating the effect of a fly ash sample on air entrainment but only for fly ash sources that demonstrate a large effect in air entrainment.

The LOI and mortar air content results of samples of all fly ash sources except C1 are plotted in Figure 5.6a. Similarly, the foam index and mortar air content results of samples of all fly ash sources except C1 are plotted in Figure 5.6b. From the plots, it is clear that a given change in value of LOI or foam index will have a different effect on mortar air content for each source. So before one can use LOI or foam index values at estimating the effects on air entrainment it is important that a correlation

TABLE 5.1
Data Analysis of Results of Fly Ash Source F1

	Initial Mortar Air, %	Final Mortar Air[a], %	LOI %	Foam Index, oz/cwt (mL/100 kg)	RD	7-Day Fly Ash Strength, psi (MPa)	7-Day SAI, %	7-Day Control Strength, psi (MPa)	% Retained No. 325	Water Demand, %
Average	10.8	8.9	1.2	2.01 (131)	2.16	3773 (26.02)	80.1	4711 (32.49)	26.5	96.6
Standard deviation	1.7	3.5	0.5	0.29 (19)	0.12	151 (1.04)	3.7	52 (0.36)	5.2	1.5
Coefficient, of variation, %	15.3	39.1	43.7	14.3	5.6	4.0	4.6	1.1	19.8	1.5
Maximum	13.4	13.9	2.1	2.40 (156)	2.44	4064 (28.03)	88.0	4791 (33.04)	38.3	100.0
Minimum	8.1	3.6	0.3	1.35 (88)	1.92	3354 (23.13)	70.0	4618 (31.85)	19.3	94.0

[a] Final mortar air as measured after 45 min (40 min of rest followed by 5 min of mixing).

[b] Foam index dosage is reported as ounces per 100 lb of cementitious or mL per 100 kg of cementitious.

TABLE 5.2
Data Analysis of Results of Fly Ash Source F2

	Initial Mortar Air, %	Final Mortar Air[a], %	LOI %	Foam Index[b], oz/cwt (mL/100 kg)	RD	7-Day Fly Ash Strength, psi (MPa)	7-Day SAI, %	7-Day Control Strength, psi (MPa)	% Retained No. 325	Water Demand, %
Average	11.8	10.5	2.0	1.58 (103)	2.17	3516 (24.25)	77.5	4538 (31.30)	23.9	98.2
Standard deviation	0.9	1.9	0.5	0.17 (11)	0.04	137 (0.94)	2.8	184 (1.27)	3.0	0.8
Coefficient, of variation, %	8.0	18.1	25.4	10.9	1.8	3.9	3.7	4.1	12.5	0.8
Maximum	13.5	13.0	3.1	1.92 (125)	2.27	3820 (26.34)	82.0	4806 (33.14)	32.0	100.0
Minimum	9.7	6.6	0.7	1.15 (75)	2.06	3214 (22.17)	68.0	4287 (29.57)	16.8	97.0

[a] Final mortar air as measured after 45 min (40 min of rest followed by 5 min of mixing).
[b] Foam index dosage is reported as ounces per 100 lb of cementitious or mL per 100 kg of cementitious.

TABLE 5.3
Data Analysis of Results of Fly Ash Source F3

	Initial Mortar Air, %	Final Mortar Air[a], %	LOI %	Foam Index, oz/cwt (mL/100 kg)[b]	RD	7-Day Fly Ash Strength, psi (MPa)	7-Day SAI, %	7-Day Control Strength, psi (MPa)	% Retained No. 325	Water Demand, %
Average	14.5	10.8	8.0	3.74 (244)	2.40	3466 (23.90)	84.7	4093 (28.23)	22.4	102.3
Standard deviation	1.8	3.6	2.1	0.91 (59)	0.06	143 (0.99)	3.4	66 (0.46)	4.3	1.7
Coefficient, of variation, %	12.5	33.7	26.3	24.3	2.3	4.1	4.0	1.6	19.4	1.6
Maximum	18.4	19.3	12.5	5.76 (376)	2.60	3956 (27.28)	97.0	4214 (29.06)	30.5	106.0
Minimum	11.1	5.2	3.6	1.54 (100)	2.32	3215 (22.17)	79.0	4034 (27.82)	13.2	99.0

[a] Final mortar air as measured after 45 min (40 min of rest followed by 5 min of mixing).
[b] Foam index dosage is reported as ounces per 100 lb of cementitious or mL per 100 kg of cementitious.

TABLE 5.4
Data Analysis of Results of Fly Ash Source C1

	Initial Mortar Air,%	Final Mortar Air[a],%	LOI %	Foam Index, oz/cwt (ml/100 kg)[b]	RD	7-Day Fly Ash Strength, psi (MPa)	7-Day SAI,%	7-Day Control Strength, psi (MPa)	% Retained No. 325	Water Demand, %
Average	14.1	14.2	0.5	1.18 (77)	2.66	4393 (30.30)	96.0	4577 (31.57)	18.1	96.4
Standard deviation	0.3	0.6	0.2	0.10 (7)	0.03	62 (0.43)	1.6	56 (0.39)	0.9	0.8
Coefficient, of variation, %	2.5	4.5	33.6	8.9	1.3	1.4	1.7	1.2	5.1	0.8
Maximum	14.6	15.3	1.2	1.54 (100)	2.73	4513 (31.12)	99.0	4653 (32.09)	19.9	98.0
Minimum	13.1	12.8	0.3	1.15 (75)	2.57	4270 (29.45)	93.0	4531 (31.25)	16.5	95.0

[a] Final mortar air as measured after 45 min (40 min of rest followed by 5 min of mixing).
[b] Foam index dosage is reported as ounces per 100 lb of cementitious or mL per 100 kg of cementitious.

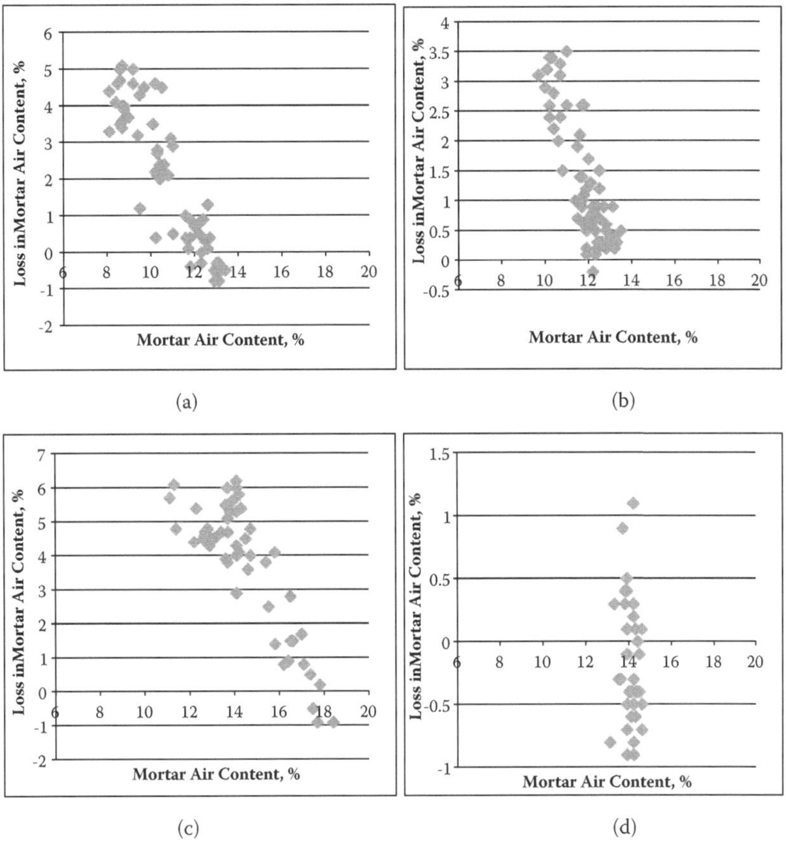

FIGURE 5.1 Increased air loss with reduced mortar air content for a fixed aea dosage (a) fly ash f1, (b) fly ash f2, (c) fly ash f3, and (d) fly ash c1.

be developed between the LOI or foam index and mortar air content for that fly ash source.

The relative color visual rating was plotted against the change in mortar air content between successive samples of each fly ash source. One would have expected the darker fly ashes to have increased reductions in mortar air content (assuming that increased carbon content resulted in the darker color). Unfortunately, no such correlations were found. Relative color does not appear to be a good way of estimating the effects of the fly ash samples on air entrainment. At best, it may indicate something has changed from the previous shipment.

Figure 5.7a–5.7d shows the relation between relative density (RD; also referred to as *specific gravity*) and mortar air content for fly ashes F1, F2, F3, and C1, respectively. Fly ashes C1 and F2 did not have good correlations between RD and mortar air content, while fly ashes F3 and F1 had better correlations. The lower RD is likely due to increase in carbon content and hence led to lower air contents.

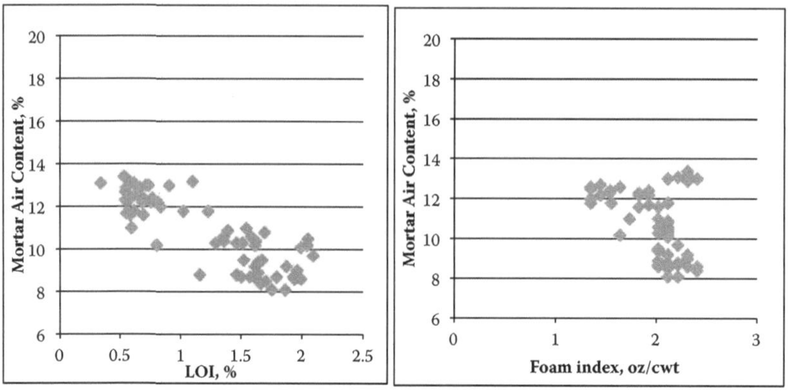

(a) (b)

FIGURE 5.2 Effect on fly ash f1 mortar air content at a fixed AEA dosage due to (a) LOI and (b) foam index, 1 oz/cwt = 65.2 mL/100 kg of cementitious.

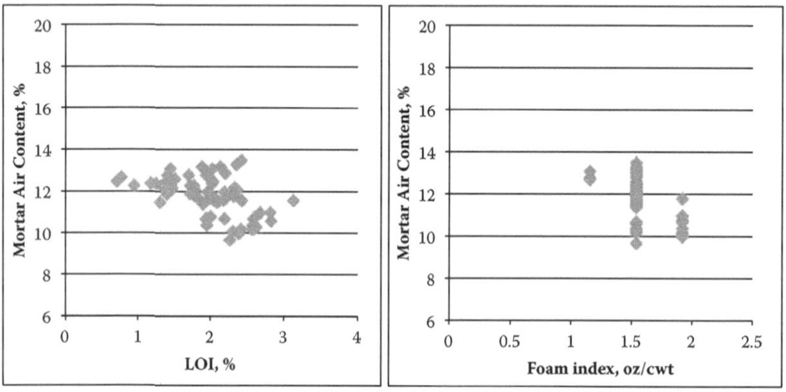

(a) (b)

FIGURE 5.3 Effect on fly ash f2 mortar air content at a fixed AEA dosage due to (a) LOI and (b) foam index, 1 oz/cwt = 65.2 mL/100 kg of cementitious.

As expected, moisture content had no correlation with mortar air content for any of the fly ashes. Fineness had an average correlation with mortar air content for fly ash F3.

STRENGTH ACTIVITY

Fly ash mortar strengths generally varied over a wider range as compared to companion control mortar strengths with the reference portland cement. One should not infer that the cement strengths are less variable because the same cement sample was used throughout the project. Therefore, the control cement mortar strength variation provided an indication of the mortar strength testing variation. The coefficient

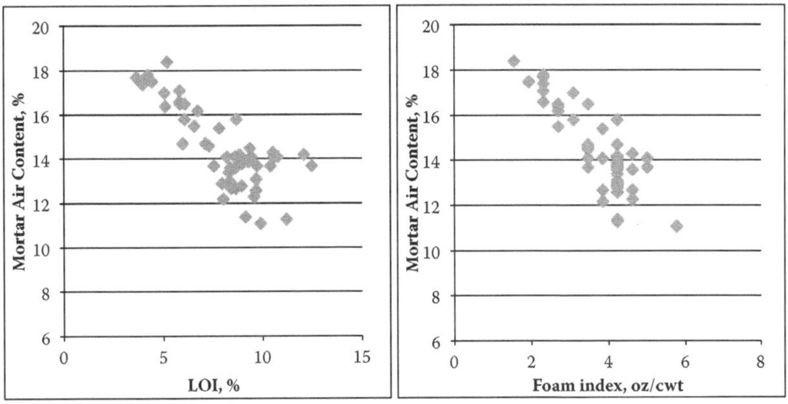

(a) (b)

FIGURE 5.4 Effect on fly ash f3 mortar air content at a fixed AEA dosage due to (a) LOI and (b) foam index, 1 oz/cwt = 65.2 mL/100 kg of cementitious.

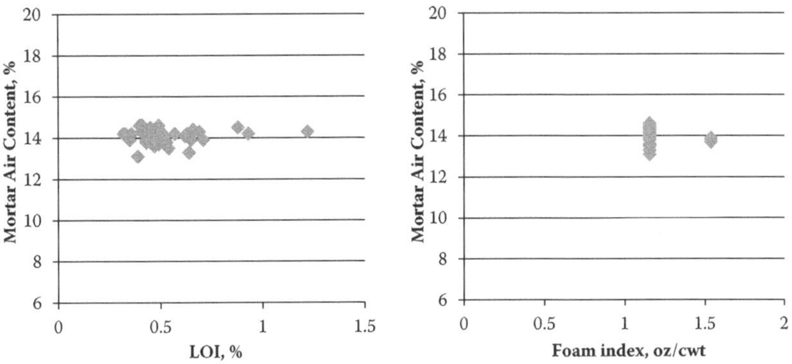

(a) (b)

FIGURE 5.5 Effect on fly ash cl mortar air content at a fixed AEA dosage due to (a) LOI and (b) foam index, 1 oz/cwt = 65.2 mL/100 kg of cementitious.

of variations (V) of the fly ash mortar strengths were about 4% for fly ash sources F1, F2, and F3. The corresponding V of the control mortar strengths of the reference cement was only about 1% with the exception of the control mortar strength tested for fly ash F2, which was inexplicably higher. The V of the fly ash mortar strengths for source C1 was also much lower at 1.4%. This suggests that the variation in strength for fly ash source C1 was largely due to testing, whereas fly ash sources F1, F2, and F3 had a statistically significant mortar strength variation, greater than that attributed to testing variation.

The average range (difference between the maximum and minimum strength attained by samples from a given fly ash source) of the 7-day fly ash mortar strengths was 682 psi (4.70 MPa) for fly sources F1, F2, and F3 and 243 psi (1.68 MPa) for

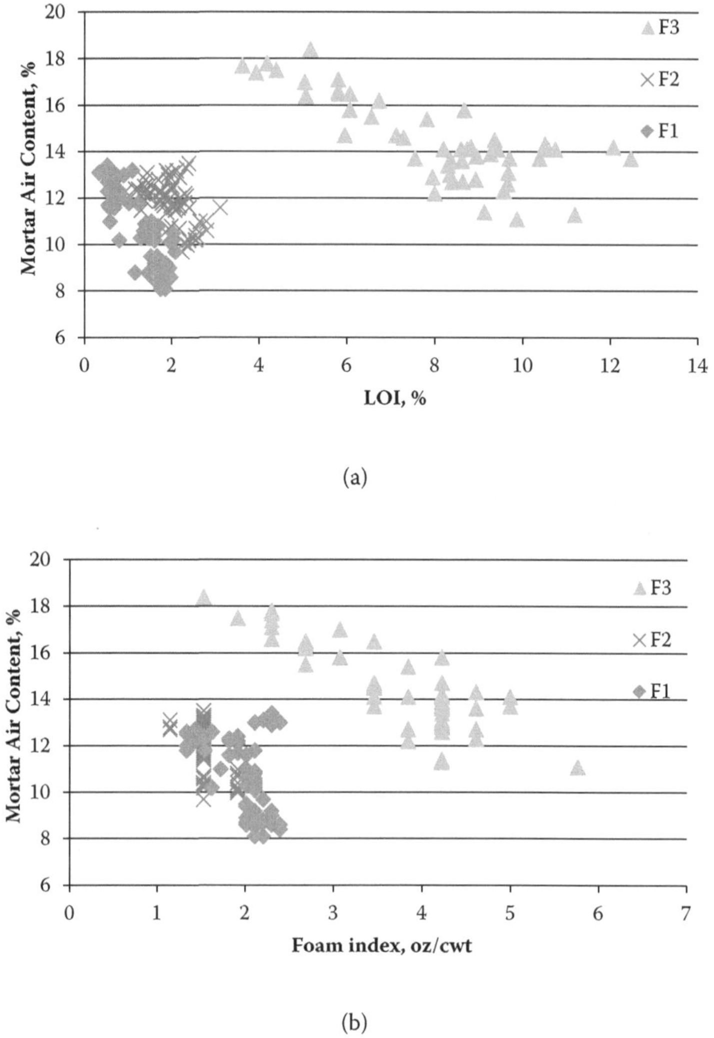

FIGURE 5.6 Effect on fly ash mortar air content at a fixed AEA dosage due to (a) LOI; and (b) foam index, 1 oz/cwt = 65.2 mL/100 kg of cementitious. Fly ashes F1, F2, F3 are plotted together.

fly ash C1. Concrete testing was not conducted in this experimental study but it would be of interest how this range of mortar strengths would translate to concrete strengths. In the earlier chapter on cement variation, it was shown that, for the same cement source, strengths of mortar cubes tested according to ASTM C109 correlated well with strengths of concrete cylinders tested according to ASTM C39. An average 1379 psi (9.51 MPa) range in C109 mortar strength (w/c = 0.485) between the

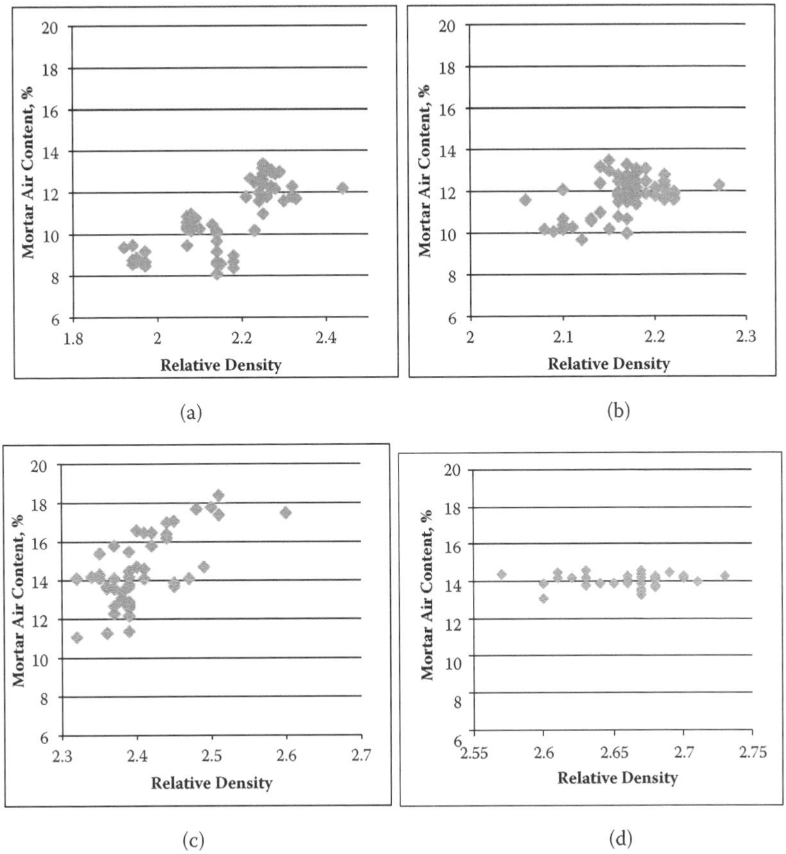

FIGURE 5.7 Effect on fly ash mortar air content at a fixed AEA dosage due to relative density: (a) fly ash F1, (b) fly ash F2, (c) fly ash F3, and (d) fly ash C1.

cement samples corresponded to an average 820 psi (5.66 MPa) range in concrete strength (w/c = 0.58) for the same samples. If a similar correlation is expected, an average range in 7-day fly ash mortar strengths of 682 psi (4.70 MPa) can correspond to an average range in concrete strength of 400 psi (2.76 MPa), as long as the same cement is used for both the mortar mixtures and the concrete mixtures.

Moisture content, LOI, RD, and color were found to have no correlation with SAI for any of the fly ashes. Figure 5.8a–d shows the relation between fineness and the 7-day SAI for fly ashes F1, F2, F3, and C1, respectively. Fly ash C1 and F2 had no correlation, whereas fly ashes F1 and F3 tended to have higher strengths with increased fineness as expected, though the correlations were still weak. Fly ash blaine fineness results (ASTM C204), if available, may be used to detect changes at the lower end of the particle size distribution, which will have a greater influence on concrete strength.

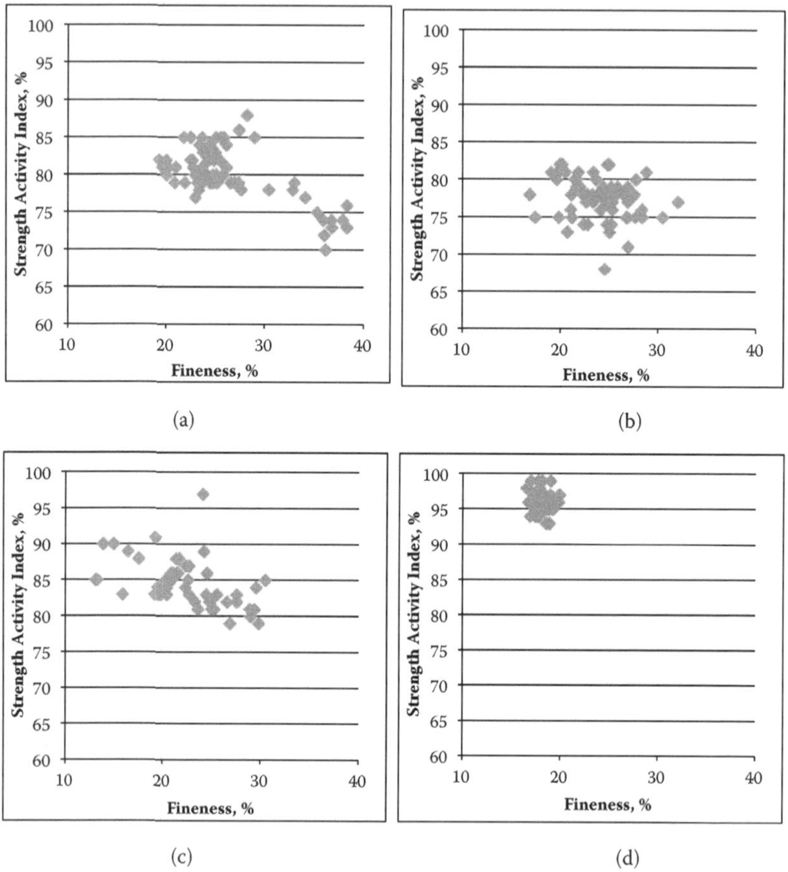

FIGURE 5.8 Effect on fly ash strength activity index due to fineness (No. 325 sieve percent retained): (a) fly ash F1, (b) fly ash F2, (c) fly ash F3, and (d) fly ash C1.

FLY ASH TESTING REQUIRED BY ASTM C311 AND C618

Table 1 in ASTM C311 provides the minimum sampling and testing frequency for fly ash. For established fly ash sources, moisture content, LOI, and fineness tests need to be conducted at a frequency of daily or every 400 tons (360 metric tons), whichever comes first; density, SAI, and various other chemical tests listed in ASTM C618 are to be conducted on composite samples monthly or every 3200 tons (2900 metric tons), whichever comes first. Composite samples do not reflect the true variation that is likely from fly ash shipments. The same composite sample test result may be applicable to many different fly ash shipments to the same concrete plant. To comply with C618, the fly ash marketer conducts LOI, and fineness testing 1–3 times a day, depending on the daily production at the fly ash source.

ASTM C618 also has a uniformity requirement for density and fineness. It states that the density and fineness of individual samples shall not vary from the average established by the 10 preceding tests, or by all preceding tests if the number is less than 10, by more than 5% (for density) and 5% retained on No. 325 sieve (for fineness).

SUGGESTED PRODUCER ACTIONS

Air Entrainment

Concrete air entrainment is perhaps the most important factor that is affected by fly ash. The concrete producer should make it a point to receive LOI and fineness test results conducted on the same day the shipment leaves the fly ash source. Sometimes the test results may be available only at the end of the day after the fly ash shipment has left for the concrete plant. It is worthwhile for the concrete producer to develop an understanding with the fly ash marketer to receive those test results on the same day the fly ash shipments are received at the concrete plant.

If the LOI test result varies from the previous fly ash shipment, that by itself is of little value. It is suggested that fly ash marketers develop a correlation between LOI and mortar air content for that fly ash source as shown in Figures 5.2 to 5.5. If the correlation between LOI and mortar air content had been found to be very poor for that fly ash source, then the fly ash marketer can develop a correlation between foam index and mortar air content. If the correlation between foam index and mortar air content is poor as well, then the fly ash marketer could conduct mortar air content testing every time the LOI test is conducted at the fly ash source. This systematic approach provides the concrete producer some understanding of the effect of the fly ash shipment on air entrainment before the fly ash is used in concrete.

The concrete producer could also conduct the foam index or the mortar air content test at the concrete plant when a fly ash shipment is received. These tests can be completed in less than 15 minutes by operators without any significant training. If the mortar air content attained is, for example, 50% less than that achieved for the previous fly ash shipment, the decision can be made to increase the AEA dosage in the concrete made with the new fly ash shipment by at least 50%. The first few concrete trucks that use the fly ash from the new shipment should be tested for air content in accordance with ASTM C231. Depending on those concrete test results AEA admixture dosages for future truckloads could be adjusted. The concrete producer could skip the foam index or mortar air content tests and conduct concrete air content testing at the plant with the new fly ash shipment while maintaining the same AEA dosage as for the previous fly ash shipment. This approach may take a little longer to clearly identify the effect of the new fly ash shipment on air entrainment because there are many other factors involved in batching and measuring air content from a concrete truck.

The RD of fly ash could also be an indicator for the effect of the fly ash on air entrainment as shown in Figure 5.7. However, RD is conducted on composite fly ash samples, and therefore, the data are attained less frequently and as discussed earlier may not be representative of the shipment received. The concrete producer should check to ensure that the fly ash meets the uniformity requirements for density. This serves a twofold purpose: (1) It ensures that the correct RD for the fly ash is used in the mixture proportions, and (2) it provides assurance that the fly ash marketer is supplying quality fly ash.

In the experimental study discussed earlier, color did not prove to be a good indicator for air entrainment. It would be useful to conduct the color test using handheld colorimeters. These are less subjective, and it would be of interest to see whether a correlation can be developed between the results of the colorimeter test and the mortar air content.

Strength Activity Index

The results of the experimental study discussed earlier suggested that different fly ash shipments from a given source can contribute to a variation in concrete strength. A concrete producer interested in reducing this strength variation can attempt to adjust the concrete mixture proportions based on the expected strength variation due to fly ash. The producer should pay close attention to the fineness and SAI values. The first thing is to ensure that the C618 uniformity requirements for fineness are met, thereby providing assurance that the fly ash marketer is supplying quality fly ash. The producer can develop a control chart of moving averages of 3 or 5 consecutive fineness test results for the shipments received at the concrete plant and adjust mixture proportions if certain control chart limits are exceeded. It is suggested that a single upper control chart limit be set at 7% above the average fineness value. When the upper fineness limit is exceeded (i.e., there is a higher percent retained), it means that strengths can be lower as the fly ash has coarser particles. It is suggested that when the upper fineness limit is exceeded, the compressive strength of concrete be increased by 150 psi (1 MPa) through the use of a lower w/cm. No lower fineness control chart limit is suggested.

SAI could also be an indicator for variation in concrete strength, but the SAI test results suffer from several disadvantages: (1) SAI tests are conducted on composite samples, (2) 7-day SAI values are likely to be available many days after the fly ash shipment has been received at the concrete plant, and (3) SAI is conducted by the fly ash marketer using their reference cement. Fly ash interaction with the cement used by the producer may be different, in which case the variation in SAI test results may not necessarily correlate with the variation in compressive strength of concrete produced at the plant. Nevertheless, it is useful to plot both the 7- and 28-day SAI values as soon they are obtained. Concrete producers can use it to troubleshoot low concrete strengths in evaluating whether the cause for the low concrete strength can be attributed to a reduction in the SAI. Other factors, such as mixing water, air content, batching errors, and testing errors, should also be considered.

Other Tests

Thermal measurements of hydrating concrete mixtures is another tool at the disposal of the concrete producer (Cost 2009). Essentially, this test involves casting a mortar or a concrete cylinder and measuring its temperature rise in an insulated environment over a 24-hour period. The temperature profile can provide indications of concrete setting time, early age strength development, and potential interaction problems between cementitious materials and admixtures. Significant variations in temperature profiles can indicate potential variation in the above properties. When these tests are done periodically at the concrete plant, they can provide a means to study the overall variations due to shipments of cement, supplementary cementitious materials, and chemical admixtures. Once a change in the overall behavior has been identified, individual material shipments could be tested to identify the root cause of the change.

SUMMARY OF SUGGESTED PRODUCER ACTIONS

Develop an understanding with the fly ash marketer so that all the LOI, fineness, foam index, and mortar air content (if available) test results conducted on the same day the fly ash shipment left the plant are attained in a timely manner.

Encourage fly ash marketers to develop a correlation between LOI and mortar air content or LOI and foam index for that fly ash source. If there is no such correlation, encourage fly ash marketers to conduct mortar air content testing every time the LOI test is conducted at the fly ash source. This systematic approach provides the concrete producer with some understanding of the effect of the fly ash shipment on air entrainment before the fly ash is used in the concrete.

Develop company policy on adjusting AEA dosage based on foam index, mortar air content, or fly ash LOI test results of the new fly ash shipment. At a minimum, test the first few concrete trucks that use the fly ash from the new shipment for air content in accordance with ASTM C231 and adjust AEA dosage accordingly. This testing can be conducted at the plant.

Check to ensure that the fly ash is meeting the uniformity requirements for density and fineness. Use the correct RD for the fly ash in concrete mixture proportions.

Develop a moving average control chart of fineness test results for the shipments received at the concrete plant. Develop company policy to adjust concrete mixture proportions if the upper control charts limit is exceeded.

Plot the 7- and 28-day SAI values as soon they are obtained. Use it to troubleshoot low concrete strength problems. Evaluate whether the cause for the low concrete strength can be attributed to a reduction of the SAI. Consider other factors, such as mixing water, air content, batching errors, and testing errors. Retain 5 lb samples of fly ash from each shipment for 3 to 6 months in sealed containers so that these can be tested at a later point if necessary. Sampling procedures for fly ash are described in ASTM C311. If possible, conduct thermal measurements of hydrating concrete mixtures and look for significant variations in temperature profiles.

6 Variation in Concrete Performance Due to Aggregates

This chapter discusses variability in concrete performance due to variation of aggregate from a single source.

VARIABILITY OF AGGREGATE FROM SINGLE SOURCE

Aggregate typically occupies about 75% of the volume or weight of a cubic yard of concrete. Even though it is largely inert, its large proportion ensures that variation in aggregate properties will have significant impact on concrete performance such as strength, water demand for a given slump, and fresh properties, such as cohesiveness, harshness, segregation, bleeding, ease of consolidation, finishability, and pumpability; all of which may not always correlate with slump.

Generally, aggregate manufacturers provide concrete producers with test data and certification that their aggregate meets the ASTM C33 Specification for Concrete Aggregates. The tests required in ASTM C33 are outlined in Table 6.1. In addition, aggregate test data on relative density (also referred to as specific gravity), absorption, and bulk density of coarse aggregate are required for concrete mixture proportioning. ACI 301 Specification for Structural Concrete states that aggregates used in the project should conform to ASTM C33, and test results showing conformance should not be older than 90 days, except for test results for soundness, abrasion, and reactivity, which should not be older than 1 year. This would require the concrete producer to have current test data of all the above aggregate tests every 90 days except for soundness, abrasion, and reactivity, which can be conducted on a yearly basis. Typically, these data are provided by the aggregate supplier.

Depending on aggregate production volumes at the quarry for internal quality control, aggregate manufacturers conduct daily or weekly testing of certain aggregate property tests. Table 6.2 adapted from ACI 221R shows a typical quality control program listing the routine control tests to be conducted by both the aggregate and the concrete producers. A smaller number of tests as compared to Table 6.1 are included here. This is due to the following reasons:

1. It is impractical to do all of the tests at the stated frequency.
2. Depending on the source, some of the aggregate properties do not change as much, and so it is adequate to do quality control tests more frequently on properties that tend to change more often.

The reader is directed to ACI 221R and ASTM STP 169D (2006) for a detailed discussion of aggregate tests and effects that the aggregates have on concrete performance. The following section briefly discusses how the aggregate test results affect concrete mixture proportioning and performance.

AGGREGATE PROPERTIES AND THEIR EFFECT ON CONCRETE MIXTURE PROPORTIONING AND PERFORMANCE

Relative Density and Absorption of Aggregate

Relative density (RD) and absorption of the aggregate tested according to ASTM C127 and C128 for coarse and fine aggregate, respectively, are unlikely to vary significantly. RD of the aggregate is used in concrete mixture proportioning, and changes in RD will change the volumetric composition of the mixture and likely result in discrepancies in yield of concrete batches. Absorption is used to calculate the batch water content of the concrete, and using incorrect values can lead to inaccurate mixing water amounts, incorrect w/cm, and therefore variations in strength and other concrete properties impacted by water content. High variation in relative density speaks to the lack of source control and will need frequent concrete mixture adjustments.

Aggregate Moisture Content

Aggregate moisture content should be measured and batchwater corrected as discussed in Chapter 4 (Mixing Water Control). An attempt should be made to maintain uniform aggregate moisture content when batching concrete. This is accomplished in fine aggregates by adopting good draining storage practices and ensuring that the fine aggregate stockpiles have been inactive long enough. While fine aggregates with a round, smooth shape can drain within 12 hours, fine aggregate that have angular/flat particles may take up to a week. While wet sand generally contributes more free moisture to a concrete batch, moisture content of coarse aggregate should not be ignored or assumed. Moisture probes that are well calibrated frequently and connected to control systems that allow for automated water adjustment can considerably improve the batch-to-batch uniformity of concrete for fresh and hardened properties.

Void Content in Coarse Aggregates

Aggregate bulk density and void content (ASTM C29), also known as the Dry Rodded Unit Weight (DRUW), is recommended for coarse aggregates. The DRUW is used in concrete mixture proportioning to establish the amount of coarse aggregate in a concrete mixture. The void content determined is a function of the aggregate particle shape, texture, and grading. If the DRUW test is conducted on a fixed grading, the void content will depend on the coarse aggregate shape and texture, with rounded aggregates with smoother texture resulting in lower void contents. The coarse aggregate void content determined in accordance with C29 in as-received grading will not differ much from that determined at a fixed grading as long as there is not an excess amount of aggregates in the finer particle sizes.

Based on an large experimental study, Bloem and Gaynor (1963) reported that when different coarse aggregate sources were used with a single source of fine aggregate to make concrete, every 1% increase in coarse aggregate void content determined according to ASTM C29 (fixed grading) led to an average increase in mixing-water content of 0.5 gal/yd^3 (2.5 L/m^3) for a concrete slump of 2 in. to 3 in. (50 mm to 75 mm). Wills (1967) tested nine gravels and found the coarse aggregate void contents determined according to ASTM C29 (fixed grading) correlated very well with the mixing-water demand (i.e., mixing-water quantity required for a fixed slump of 3 in. to 4 in. [75 mm to 100 mm]). The void contents varied from 33% to 42%, and the corresponding mixing-water demand for concrete with a control fine aggregate ranged by about 33 lb/yd^3 (20 kg/m^3). Large amounts of flat and elongated particles as measured according to ASTM D4791 can make concrete mixtures too harsh for some placement methods, resulting in voids, honeycombing, or pump blockages.

Void Content of Fine Aggregates

ASTM C1252 describes the determination of the uncompacted void content of fine aggregate. If the test is conducted on an as-received aggregate grading (Method C), the void content is influenced by particle shape, surface texture, and grading. If the test is conducted on a standard aggregate grading (Method A), the void content depends on the aggregate shape and texture, with rounded aggregates with smoother texture giving lower void contents.

Wills (1967) tested nine fine aggregates and found that the void content measured using a fixed grading—similar to ASTM C1252 (Method A)—correlated very well with the mixing-water demand. The void contents of fine aggregates varied from 39% to 50%; the corresponding mixing-water demand for concrete with a control gravel for a fixed slump of 3 in. to 4 in. (75 mm to 100 mm) ranged by about 50 lb/yd^3 (30 kg/m^3; Figure 6.1), and the compressive strength ranged by about 2000 psi (13.8 MPa; Figure 6.2). Bloem and Gaynor (1963) and Wills (1967) concluded that fine aggregate particle shape and texture had a greater influence on the mixing-water content and compressive strength than coarse aggregate particle shape and texture.

After reviewing various studies, Gaynor and Meininger (1983) reported that every 1% increase in fine aggregate void content (measured at a fixed grading) resulted in an increase in mixing-water content of 3 to 8 lb/yd^3 (2 to 5 kg/m^3) for a target slump; the higher value would apply when different aggregate sources are compared and the smaller value would apply when changes in processing change particle shape and/or texture at a single source.

Aggregate Grading

The aggregate sieve analysis, conducted in accordance with ASTM C136 and the fineness modulus (FM), must be determined for both coarse and fine aggregates. The fineness modulus (FM) is an empirical number related to the aggregate grading with higher FMs corresponding to coarser aggregates. However, aggregates with the same FM can have different grading. Even though FM can be calculated for all aggregates, typically it is used to characterize the grading of only fine aggregates. ASTM C33

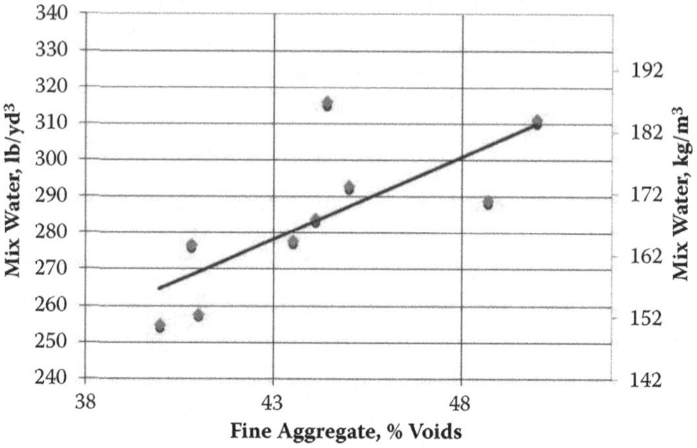

FIGURE 6.1 Mixing-water demand for concrete made with different fine aggregates and control gravel. (Reprinted with permission from Wills, M. H., Jr., "How Aggregate Particle Shape Influences Concrete Mixing Water Requirement and Strength," *Journal of Materials*, Published by ASTM, Vol. 2, No. 4, December 1967, pp. 843–865.)

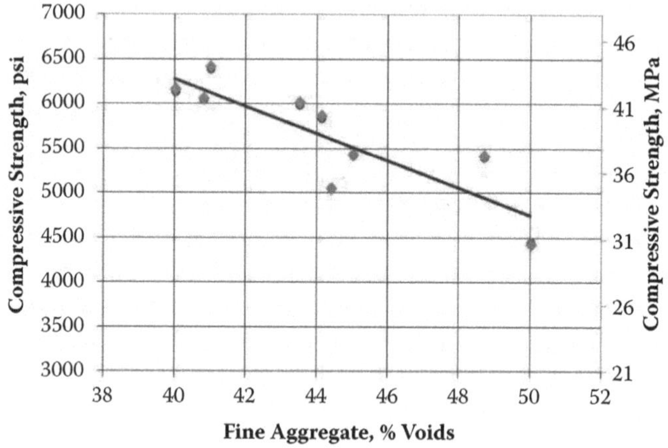

FIGURE 6.2 Compressive strength of concrete made with different fine aggregates and control gravel. (Reprinted with permission from Wills, M. H., Jr., "How Aggregate Particle Shape Influences Concrete Mixing Water Requirement and Strength," *Journal of Materials*, Published by ASTM, Vol. 2, No. 4, December 1967, pp. 843–865.)

has requirements on the grading of coarse and fine aggregates and places limits on the FM of fine aggregates. ASTM C33 requires concrete fine aggregate to have an FM between 2.3 and 3.1. As a control on same source uniformity, it indicates that the FM should not vary by more than 0.20 from the base FM. In the ACI 211.1 mixture proportioning procedure, the FM of the fine aggregate is used in conjunction with the DRUW of the coarse aggregate to determine the aggregate proportions in concrete mixtures. Finer sands (lower FM) result in increased coarse aggregate content at the same nominal maximum size of coarse aggregates. Changes in fine aggregate FM over a range as much as 1.0 have been noted in a day's production (ASTM 169D). If the fine aggregate FM changes more than 0.20, then the relative proportions of coarse and fine aggregate amounts must be adjusted, as recommended by ACI 211.1. An increase in the sand FM by 0.20 will reduce the coarse aggregate quantity by about 60 lb/yd^3 (36 kg/m^3) with a similar increase in the weight of fine aggregate. Another option is to change the coarse and fine aggregate amounts such that the FM of the combined aggregate stays the same, even though the FM of the coarse or fine aggregate has changed.

Fine aggregate grading has a much greater effect on workability of concrete than does coarse aggregate grading (ASTM STP 169D). Apart from FM, it is useful to track the percentage retained on each (or selected) sieve size, particularly for the fine aggregate, on a 3- or 5-test moving average control chart for the following reasons:

1. In fine aggregate, the amount of material passing the 300 μm (No. 50) sieve should be 15% to 30% for good pumpability (ASTM 169D).
2. Fine aggregate particles between 600 μm and 150 μm entrap more air than either finer or coarser particles (Kosmatka et al. 2002). Higher amounts of material finer than 150 μm (No. 100) sieve size will cause a significant reduction in air content.
3. The amount of fine aggregate passing the 300 μm (No. 50) and 150 μm (No. 100) sieve has a great influence on workability, finishability, stickiness, potential for segregation, and bleeding of concrete (Kosmatka et al. 2002). The total amount of fines that includes the cementitious materials and that forms aggregates should be evaluated when workability is adversely effected. Hand-finishing may require higher fines content as opposed to machine finishing.

ACI 304R provides good practices for handling aggregates at the concrete plant to prevent segregation, contamination, variation in moisture content, and degradation, resulting in more fines. Some of the important recommendations are as follows:

- Build stockpiles, where necessary, in horizontal or gently sloping layers. Avoid conical stockpiles or any unloading procedure involving dumping of aggregates downsloping sides of piles.
- Keep trucks, bulldozers, and wheel loaders off stockpiles to prevent degradation and contamination.

- Prevent overlap of the different aggregate sizes by suitable walls or ample spacing between piles.
- Protect dry fine aggregate by the wind, using tarps or windbreaks.
- Fine aggregate transported over wet, unimproved haul roads can become contaminated with clay lumps, which usually accumulate between the tires and on mud flaps and gets dislodged during dumping of the transportation unit; to remove this, place a scalping screen over the batch plant bin.
- If possible, separate aggregates into individual sizes and batch separately to minimize segregation
- If aggregate degradation is likely, rescreen coarse aggregate as it is charged to the bins at the concrete plant to maintain undersized materials (minus No. 4 sieve) to as low as 2%.
- Aggregate stockpiles should be built on a base that minimizes ground contamination when aggregates are removed for intra-plant movement.

Material Finer than 75 μm (No. 200)

Aggregate particles that are finer than the 75 μm sieve (No. 200) are measured by ASTM C117 and are generally composed of silt and clay for natural sands or gravels (ASTM C169D). For crushed stone or manufactured sand, the fines are predominantly composed of dust of fracture. ASTM C33 has limits on material finer than 75 μm (No. 200) sieve for both coarse and fine aggregates. A higher amount of material finer than the No. 200 sieve will typically result in increased water demand and reduced air content. This is more significant when the fines are largely composed of clay and/or shale as opposed to being primarily dust of fracture from crushing. A variation on the material finer than the No. 200 sieve can result in variation in water demand and air content. The variation on the material finer than No. 200 sieve can occur due to the following reasons:

1. Poor quality control in washing of aggregates at the pit or quarry, which can result in dirty aggregates with coatings.
2. Aggregates stockpiled in a wet condition can collect wind-borne dust and passing traffic.
3. Improper practices by loader operator when stockpiling and removing materials from stockpiles.

Aggregate surface coatings can also prevent the cement paste from adhering to the aggregate particle, causing strength problems and may chemically react and stain the concrete.

In mixtures with higher cementitious contents, the cementitious fines tend to provide cohesion, and the presence of higher amounts of aggregate material finer than No. 200 sieve can lead to further increase in stickiness and related workability problems. In lower cementitious-content mixtures, workability and cohesion can be improved by the presence of higher amounts of aggregate material passing the No. 200 sieve. Self-consolidating concrete has been successfully developed using mineral filler (fines) to improve the stability (resistance to segregation) of fresh concrete. So, a decrease in the material passing the No. 200 may not always lead to improved workability. What is clear is it will lead to a change in workability.

Sand Equivalency

The sand equivalent test (ASTM D2419) indicates the relative proportions of clay-like or plastic fines and dust in fine aggregate that pass the 4.75 mm (No. 4) sieve. Specifications typically require a minimum sand equivalency (SE) of 70 or 75 in this test. A low SE value suggests more clay-like or plastic fines, and therefore the water demand and fresh properties of concrete can be affected. After testing about 150 fine aggregate sources, Gaynor and Meininger (1983) reported that several aggregates passing the ASTM C33 limits on material finer than No. 200 sieve still failed the typical limits set for the SE test. A methylene blue test (AASHTO T330) may be an even better indicator of the presence of clay in the aggregate. Different forms of clay minerals absorb methylene blue indicator to different degrees, so the type of clay in aggregates needs to be known to use this test.

USING AGGREGATE TEST RESULTS

Concrete producers should develop relationships with aggregate manufacturers so that the tests required by ASTM C33 and ACI 301 (Table 6.1) and the tests conducted by aggregate manufacturers for quality control (Table 6.2) are received at the stated frequency. ACI 221R states that the aggregate producer should test the material as it is loaded out of the aggregate producer's stockpiles to be shipped to the customer. The concrete producer then assumes responsibility for grading variations generated between the point of materials load-out and use in concrete. Concrete producers should take aggregate samples for testing as close to the batching process as possible, understanding that aggregate grading changes every time it is moved.

Table 6.1 Test Results

The first thing is to ensure that the aggregates that are supplied meet ASTM C33 requirements as required by ACI 301. Most of the test results are required every 90 days, except for soundness, abrasion, and reactivity, which are required on a yearly basis. As soon as the test results are available, the concrete producer should ensure that the test results meet the requirements of ASTM C33.

Table 6.2 Test Results—Tests Conducted by the Aggregate Producer

The aggregate producer should carefully look at the grading and FM results to ensure that the percent passing each sieve meets grading limits set in ASTM C33. It may be acceptable to have one in five consecutive grading test results fall outside the limits. Moving average of 5 test results shows trends in the grading results not otherwise apparent. The aggregate producer can use that to adjust the aggregate plant

TABLE 6.1
Aggregates Tests Required in ASTM C33 and ACI 301

Aggregate Type	Required Tests
Both coarse and fine aggregates	Grading and Fineness Modulus (ASTM C136), amount of material finer than 0.075 mm (No. 200) sieve (C117), coal and lignite (C123), chert (C123 and, if necessary, C295), clay lumps, and friable particles (C142), sulfate soundness[a] (C88), alkali silica reactivity[a] (C1260, and C1293)
Coarse only	Abrasion[a] (C131 or C535)
Fine only	Organic impurities (C40 and, if necessary, C87)

Note: All aggregates for the tests should be sampled according to ASTM D75 and, if necessary, the sample size reduced according to ASTM C702. ASTM C33 has limits on all of the tests. For chert, there is limit only on coarse aggregate. For fineness modulus, there is limit only on fine aggregate.

[a] ACI 301 requires that results of these tests should not be older than 1 year. Results of all other tests should not be older than 90 days.

parameters to maintain a certain target value. If the grading is controlled, the FM is likely to vary in a narrow range (±0.2 from the base FM) as required by ASTM C33 for fine aggregate. The aggregate producer should look at the void contents of the coarse and fine aggregates, which are primarily influenced by aggregate shape and texture. If the results fluctuate by more than plus or minus 2 percent from the base value, then the producer should undertake process changes to ensure that the shape and texture does not vary significantly. On the contrary, the aggregate producer can use control charts of moving average of 3 or 5 void content test results to adjust the aggregate plant to maintain a target void content. The aggregate producer can use similar 3 or 5 test moving average control charts for material finer than 75 μm (No. 200 sieve) test results to adjust the aggregate plant so that the material stays within ASTM C33 limits. SE test results are valid for certain fine aggregate sources and when required to be tested by local authorities, as discussed earlier. For those sources, the aggregate producer can use 3 or 5 test moving average control charts for SE test results to adjust the aggregate plant so that the material consistently has SE values over 70%.

The concrete producer should make appropriate adjustments to concrete mixture proportions based on the relative density, absorption and dry rodded unit weight of coarse aggregate and grading and FM of sand results that the aggregate producer supplies every 90 days. In addition, the concrete producer should carefully look at the grading, FM, void content, minus 75 μm material, and SE test results, and make sure that the aggregate plant is making efforts to produce a material consistent enough to help attain low variability concrete.

Table 6.2 Test Results—Tests Conducted by the Concrete Producer

The concrete producer establishes credibility of the test reports from each aggregate source by conducting verification tests on relative density and absorption (possibly

TABLE 6.2
Suggested Quality Control Program for Aggregates

Test	Test Method	Minimum Test Frequency
	Aggregate Plant Samples	
Coarse Aggregate		
Grading and FM	ASTM C136	Once per day
Finer than 75 μm (No. 200 sieve)	ASTM C117	Once per day
Void content	ASTM C29	Once per week
Relative density and absorption	ASTM C127	Once per 90 days
Fine aggregate		
Grading and FM	ASTM C136	Once per day
Finer than 75 μm (No. 200 sieve)	ASTM C117	Once per day
Void content (fixed grading)	ASTM C1252 (Method A)	Twice per week
Relative density and absorption	ASTM C128	Once per 90 days
Sand equivalency	ASTM D2419	Twice per week[a]
	Concrete Plant Samples	
Coarse Aggregate		
Grading and FM	ASTM C136	Once per 2 weeks
Finer than 75 μm (No. 200 sieve)	ASTM C117	Once per week
Void content (fixed grading)	ASTM C29	Once per 2 weeks
Relative density and absorption	ASTM C127	Once per year
Fine Aggregate		
Grading and FM	ASTM C136	Once per week
Finer than 75 μm (No. 200 sieve)	ASTM C117	Once per week
Void content (fixed grading)	ASTM C1252 (Method A)	Once per week
Sand equivalency	ASTM D2419	Once per week[a]
Relative density and absorption	ASTM C128	Once per year

[a] Recommended only for those aggregates known to have a problem.
Aggregate moisture testing is discussed in an earlier chapter.

other properties) on an annual basis. The results should be sent to the aggregate producer and significant variations (if any) should be discussed.

The fine aggregate grading and void content tests are conducted at a higher frequency than that for coarse aggregates to reflect the greater influence fine aggregate grading, shape, and texture has on concrete performance. Void content tests for both coarse and fine aggregates should be conducted at a fixed grading to isolate the effect of aggregate shape and texture from that of grading. There are no specification requirements for void content. If the test results between consecutive shipments vary by more than 2.0%, then the concrete mixing-water demand may change as discussed earlier, and therefore suitable changes to water-reducing admixture dosages may be needed. A moving average of five test results shows trends in the results not otherwise apparent. If there is a noticeable trend or if the results fluctuate by more than ±1.0% from the base value, the producer can share the results with the aggregate producer and have discussions to ensure that the aggregate shape and texture do not vary substantially.

If the grading test results show a greater variation than that obtained from the aggregate producer, it is clear that the stockpiling and handling practices at the concrete plant may have to be improved. If the grading test results exceed ASTM C33 requirements, then the results should be shared with the aggregate producer and a check test made. It may be acceptable to have one in five consecutive grading test results fall outside the limits. If the material consistently falls outside the specification limits, it should not be used in concrete manufacture, and the cause of the problem should be investigated. As discussed earlier, the percent passing the finer sieves for the fine aggregates can significantly influence the concrete performance. So, if significant changes are noted in those sieve sizes, concrete mixture adjustments (if allowed) can be made as discussed earlier. To get an overall effect of grading, FM values can be used. If the fine aggregate FM changes more than 0.20, then the coarse aggregate and fine aggregate amounts must be changed as recommended by ACI 211.1.

A –75 μm (No. 200 sieve) test result should be plotted on a moving average control chart. If the results exceed ASTM C33 requirements, then the results should be shared with the aggregate producer and a check test made. Stockpiling and handling practices should also be scrutinized to see whether fines are inadvertently being incorporated. If the test results between consecutive shipments vary by more than 1.0%, then concrete performance can be influenced as discussed earlier. Suitable adjustments to concrete mixture proportions may be made if allowed.

SE test results should be plotted on a moving average control chart. If the results decrease below 75%, the aggregate producer should be asked to take efforts to increase the results. If the results decrease below 70%, a different fine aggregate source may have to be considered until this source increases above 70%.

7 Basic Statistics

BASIC STATISTICAL PARAMETERS

VARIABILITY

Materials, production processes, and testing procedures are variable by nature. It is important to quantify variability with some measures. Range, R, is a simple measure of variability. It is the difference between the highest and the lowest values of the results. Standard deviation, S, is the most commonly used measure of variability. The difference (or deviation) of each result from the average are squared and summed up. This sum is divided by one less the total number of results (n − 1), and the square root of that quantity gives the standard deviation.

Mathematically, the standard deviation is expressed as follows:

$$S = \sqrt{\frac{\sum (X_i - \overline{X})^2}{(n-1)}},$$

where:

S = Standard deviation
Σ means "to take the sum of"
X_i = The individual test result
$\overline{X}$ = Average
n = Number of test results

Table 7.1 shows an example calculation of standard deviation for a list of numbers.

Other equivalent formulae that make for easier calculation of standard deviation are

$$S = \sqrt{\frac{\sum X_i^2 - \frac{\left(\sum X_i^2\right)}{n}}{(n-1)}} \quad \text{or} \quad S = \sqrt{\frac{\sum X_i^2 - n\overline{X}^2}{(n-1)}}$$

While the calculation is relatively cumbersome, standard deviation is easily calculated using calculators with statistical functions or computer spreadsheets.

In most cases, the sample standard deviation is calculated using the formula shown. The sample standard deviation is an *estimate* of the standard deviation of

TABLE 7.1
Calculation of Standard Deviation

X_i	Average $\bar{X}$	Deviations $(X_i - \bar{X})$		Deviations Squared $(X_i - \bar{X})^2$
3100		3100–3500	–400	160,000
3920		3920–3500	+420	176,400
3480	3500	3480–3500	–20	400
3700		3700–3500	+200	40,000
2950		2950–3500	–550	302,500
3850		3850–3500	+530	122,500
			Total	801,800

$$S = \sqrt{\frac{801{,}800}{(6-1)}} = \sqrt{160{,}360} = 400$$

the population. If we have results for the entire population, we would calculate the population standard deviation, in which case we divide by "n" instead of "(n – 1)." Calculators and spreadsheets have different keys or formulas, respectively, to calculate the appropriate standard deviation. The difference between "population" and "sample" standard deviation tends to get small when the number of samples, n, is larger than about 15.

For most test data, the "sample standard deviation," typically denoted by S in this book, is calculated.

Coefficient of variation, V, is the standard deviation expressed as a percentage of the average.

The mathematical formula is as follows:

$$V = \left(\frac{S}{\bar{X}}\right) \times 100$$

For the data above,

$$V = \left(\frac{400}{3500}\right) \times 100 = 11.4\%$$

Frequency Distributions

A frequency distribution is a useful way to view test data. The frequency distribution is also the basis for statistical reasoning and predictions. It forms the basis of the "normal distribution" from which ACI equations for calculating required average strength, percent within limits (PWL) concepts, and control charts are derived.

Data from a particular process will vary over a certain range. The data are arranged in ascending order. A frequency distribution is developed by grouping and

TABLE 7.2
Development of A Frequency Distribution

Range	Range Midpoint	Number of Results
2100–2299	2200	0
2300–2499	2400	1
2500–2699	2600	4
2700–2899	2800	8
2900–3099	3000	7
3100–3299	3200	12
3300–3499	3400	16
3500–3699	3600	18
3700–3899	3800	14
3900–4099	4000	9
4100–4299	4200	7
4300–4499	4400	3
4500–4699	4600	1

The following statistical parameters are determined from the results:
Number of tests, n = 100
Average, $\bar{X}$ = 3500 psi (24.14 MPa)
Standard deviation, S = 474 psi (3.27 MPa)
Coefficient of variation, V = 13.5%

counting the data in different ranges or classes. For example, for a given data set of 100, the data are grouped in ranges of 200 as shown in Table 7.2. A plot of the ranges (range midpoint) on the X-axis and the number of tests falling in each range on the Y-axis is called a *frequency histogram*. The frequency histogram for the data grouped in Table 7.2 is shown in Figure 7.1. At a glance, the frequency histogram reveals the **center** (or approximate average), the **extreme limits**, and the **spread** (or dispersion) of the data.

Normal Distribution

When a large amount of data are plotted on a frequency histogram, the data follow a typical pattern. For most data, the distribution can be approximated by a *normal frequency distribution curve*, often called a "bell curve." Figure 7.2 shows the normal frequency distribution curve drawn for the Table 7.2 data.

The normal distribution curve has several interesting properties:

- The curve is symmetrical about the average, i.e., 50% of the tests will be on either side of the average.
- The peak of the curve occurs at the average of the data.
- The spread of the curve is characterized by the standard deviation. The greater the spread, the higher the standard deviation.
- Statistical theory can be used to predict the percentage of tests within (or outside) a particular range. If the total area under the curve represents 100% of the results, a portion of that area is proportional to the number of tests within that range.

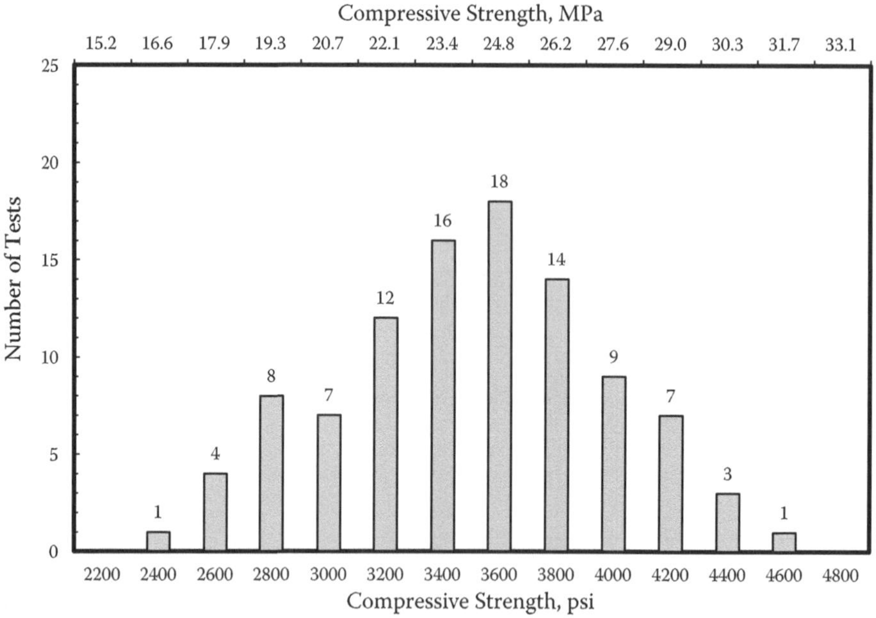

FIGURE 7.1 Frequency histogram of 100 strength test results.

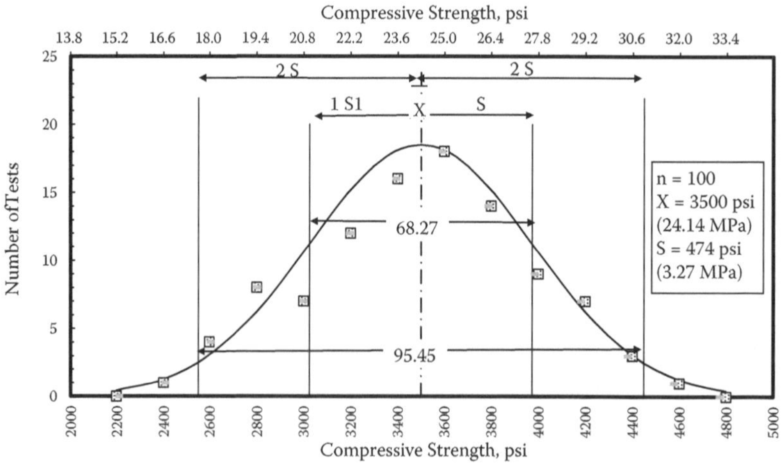

FIGURE 7.2 Normal distribution curve for strength test results.

For example, statistical theory tells us that for a normal distribution, 68.27% of the area under the bell curve will be in the range between ± 1 standard deviation about the average, that is, between ($\bar{X}$ – 1 S) and ($\bar{X}$ + 1 S). This indicates that 68.27% of the tests results will fall in this range, as illustrated in Figure 7.2.

In the example above, about 68 of the 100 test results will be between

$$(\bar{X} - 1\ S) = (3500 - 474) = 3026 \text{ psi } (24.1 - 3.27 = 20.87 \text{ MPa})$$

and

$$(\bar{X} + 1\ S) = (3500 + 474) = 3974 \text{ psi } (24.1 + 3.27 = 27.37 \text{ MPa})$$

It follows that the remaining 32 tests will be outside this range. As the distribution is symmetrical about the average, one half, or 16 tests, will be greater than 3974 psi (27.37 MPa), and 16 tests will be less than 3026 psi (20.87 MPa).

Similarly, 95.45% of the data will fall between ($\bar{X}$ – 2 S) = 2551 psi (17.59 MPa) and ($\bar{X}$ + 2 S) = 4449 psi (30.68 MPa), and 4.55% of the tests will fall outside that range.

Predictions Using a Normal Distribution

If the average and standard deviation of a set of data are known, then by assuming a *normal distribution*, the probability (or odds) of a test falling within a certain range, expressed as the average plus or minus some fraction of the standard deviation can be predicted. The expected percent of tests *within* a given function of the average, $\bar{X}$, and the standard deviation, S, can be predicted from Table 7.3 (percentages have been rounded off).

TABLE 7.3
Percentage of Tests within Indicated Ranges of the Normal Distribution

Strength Level	Percent
$\bar{X}$ ± 0.43 S	33
$\bar{X}$ ± 0.84 S	60
$\bar{X}$ ± 1.00 S	68
$\bar{X}$ ± 1.28 S	80
$\bar{X}$ ± 1.64 S	90
$\bar{X}$ ± 1.96 S	95
$\bar{X}$ ± 2.00 S	95.5
$\bar{X}$ ± 3.00 S	99.8

Example: Based on 50 sieve analysis tests on a concrete sand, we have an average of 25.0% passing the No. 50 sieve and a standard deviation of 3.9%. If the specification limits are 20 to 30% passing, what percent of tests meet the specification?

The lower specification limit = (25–20) ÷ 3.9 = 1.28 standard deviations below the average.
The upper specification limit = (30–25) ÷ 3.9 = 1.28 standard deviations above the average.

Assuming the data follows a normal distribution, Table 7.3 indicates that 80% of the tests (40 tests) will fall in the range of $\bar{X} \pm 1.28$ S and thereby conform with the specification.

This works quite well provided the assumption that the data following a normal distribution is accurate, and the statistics are calculated with a large amount of data (at least 30 tests).

The number of tests that will be less than a certain value can be predicted by expressing that value as the average minus some factor of the standard deviation. The expected percent of tests *below* a given function of the average, $\bar{X}$, and the standard deviation, S, can be predicted from Table 7.4.

Example: The specified strength for a particular class of concrete is 3000 psi (20 MPa). The standard deviation, calculated from 30 tests of this class of concrete, is 450 psi (3 MPa). What should the average strength of the concrete be if the specification indicates that it is acceptable to have 1 in 20 (5%) tests less than the specified strength.

The mixture should be designed to produce an average strength high enough such that, assuming a normal distribution, only 5% of the tests will be less than 3000 psi (20 MPa).

The average strength should be 3000 + (**1.64** × 450) = 3738 psi or 20 + (**1.64** × 3) = 24.9 MPa.

TABLE 7.4
Percentage of Tests Less Than Indicated Value of the Normal Distribution

	Below Value	
Strength Level	**Percent**	**Number**
$\bar{X}$ – 0.43 S	33	1 in 3
$\bar{X}$ – 0.84 S	20	1 in 5
$\bar{X}$ – 1.00 S	16	1 in 6.25
$\bar{X}$ – 1.28 S	10	1 in 10
$\bar{X}$ – 1.64 S	5	1 in 20
$\bar{X}$ – 1.96 S	2.5	1 in 40
$\bar{X}$ – 2.00 S	2.25	1 in 44
$\bar{X}$ – 3.00 S	0.10	1 in 1000

This forms the basis of the ACI equations to calculate the required overdesign strength for a specified strength when the standard deviations of strength test results of concrete produced from a certain plant are available. ACI 318 requirements are covered later in this section, and in Appendix A.

Types of Variation

Common Causes and Special Causes

The variability of any manufactured product is unavoidable. The variation may be attributed to chance causes, which nowadays are called *common causes,* or assignable causes, which are now called *special causes.*

Common causes are attributed to the normal variability of the process. Common causes are typically chronic and usually are due to multiple minor variables. They are difficult to diagnose and provide remedies.

For example, the slight variation in batching equipment may be attributed to a common cause.

Special causes are typically sporadic and often are due to single variables. Special causes may be identified by studying one or more control charts.

For example, if the batching equipment starts malfunctioning, it may be attributed to a special cause. Similarly, a series of low-strength tests associated with a new inexperienced testing operator is a special cause.

Step Changes

Sometimes the variations in a property (strength, for example) are sustained, and this results in a change in average value. This sudden change in average strength is called a *step change.* Step changes could be due to changed constituent materials being used, weigh-scales and testing machines that malfunction with a consistent bias, and so forth. With concrete production, step changes in average strength are more common than drifts in average strength (Gibb and Harrison 2010). An example of a step change is provided in a later section on CUSUM charts.

Control Charts

One of the important tools in quality control is a control chart. Control charts depict process changes over time. They were first developed by Walter A. Shewhart, and so they are also called Shewhart control charts. It is important to note that control charts do not control the process but are a means of verifying that the process is in control and making changes if it is not.

Control charts can be useful to distinguish special causes from common causes. It is also important that a common cause not be misinterpreted as a special cause, in which case, the reason for the variability will be difficult to establish.

A control chart can be viewed as a distribution turned sideways with the vertical axis being the test results and the horizontal axis being the successive test numbers.

Some of the benefits of using control charts include:

- Detecting trouble early
- Decreasing variability

- Establishing the process capability
- Reducing price adjustment costs
- Decreasing inspection frequency
- Using as a basis for changing the specification limits
- Permanent record of quality
- Instilling quality awareness

Individual Chart

The simplest type of control chart is the individual chart. The individual test results are plotted and checked against some control limits. The control limits might be

- Specification tolerances plotted about a target value. In this case, action should be taken before the data approaches the specification limits to avoid rejection of the product.
- Arbitrary control limits plotted about a target value. The control limits are selected so that corrective action may be taken when the data falls outside their bounds. Arbitrary control limits should be tighter than the specification tolerances. For example, if the specification calls for air content at 6 ± 1.5%, the control limits may be set at arbitrary limits of ±1%.
- Control limits based on the standard deviation calculated from historical (recent) data plotted about the average. The previous data should include at least 15, and preferably 30, data points. The control limits may be set at ±2 (or 3) standard deviations. Generally, control limits are set at ±3 S. If the data points lie within the control limits, the process is said to be in statistical control. Points that fall outside the control limits are said to come from special causes. So, in this respect, the charts provide a decision-making tool to identify when a special cause for the variation occurs that then generates an action to correct the situation or not. A lower factor of the standard deviation might be preferable to generate a faster response to a potential change where corrective action may be required. For example, if the control limits are set at ±2 S, then a change in the process might be identified when more than approximately 1 in 20 tests (5%) fall outside the control limits. It is also worth noting that there is only a 0.2% probability that two consecutive test results fall outside the 28-day or 7-day control limits set at ±2S and 0.8% probability that seven consecutive test results are on one side of the average.

The control limits on a control chart are not to be used to determine whether the product is acceptable but to indicate that there is a change in the process and generate corrective actions.

Average and Range Charts

The control chart for average (X-bar chart) and the control chart for ranges (R-chart) are used together to identify process changes. The X-bar chart is useful to detect when the process target or average changes and the R-chart is used to determine when the process variability changes.

The advantage of a control chart for average (X-bar chart) is that, even if individual values of a process do not follow a "normal distribution," the distribution of the averages will be approximately normal. Control limits of ($\pm 3S_{\bar{X}}$) are typically used to identify changes in the process average. $S_{\bar{X}}$ is the standard deviation of the averages of each subgroup of data. The control limits for average control charts should be tighter than that for control charts of individual results. The standard deviation, calculated from individual results, should be modified by dividing by $\sqrt{n}$, for charts of averages, where n is the number of individual results being averaged.

Example: The standard deviation, S, of 30 tests is 300 psi (2.07 MPa), and the control limits are chosen as ±1.64 S.

The control limits for the control chart plotting individual test results

$$= \bar{X} \pm (1.64 \times 300) = \bar{X} \pm 492 \text{ psi or } \bar{X} \pm (1.64 \times 2.07) = \bar{X} \pm 3.4 \text{ MPa}$$

The control limits for the control chart of an average of three tests

$$= \bar{X} \pm \left[1.64 \times \left(300 \div \sqrt{3}\right)\right] = \bar{X} \pm 284 \text{ psi or } \bar{X} \pm \left[1.64 \times \left(2.07 \div \sqrt{3}\right)\right]$$

$$= \bar{X} \pm 1.96 \text{ MPa}$$

More than one test result (n > 1) in each subgroup is required to calculate the ranges for the R-chart. However, generally, the standard deviation need not be calculated, and the following approach can be used:

The control limits for the X-bar chart are calculated as follows:

Upper Control Limit

$$UCL_{\bar{X}} = \bar{\bar{X}} + (A_2 \times \bar{R}) \text{ (used to estimate } \bar{\bar{X}} + 3\ S_{\bar{X}})$$

Lower Control Limit

$$LCL_{\bar{X}} = \bar{\bar{X}} - (A_2 \times \bar{R}) \text{ (used to estimate } \bar{\bar{X}} - 3\ S_{\bar{X}})$$

The control limits for the R-chart are calculated as follows:

Upper Control Limit

$$UCL_R = D_4 \times \bar{R} \text{ (used to estimate } \bar{R} + 3\ S_R)$$

TABLE 7.5
Factors for Control Limits of Statistical Control Charts

Tests in	Average	Range	
Subgroup, n	A_2	D_3	D_4
2	1.88	0	3.27
3	1.02	0	2.58
4	0.73	0	2.28
5	0.35	0	2.12
6	0.48	0	2.00
7	0.42	0.08	1.92

Lower Control Limit

$$LCL_R = D_3 \times \overline{R} \text{ (used to estimate } \overline{R} - 3\ S_R),$$

where

$\overline{\overline{X}}$ = overall average, i.e., average of sample averages
$\overline{R}$ = Average range, i.e., average of sample ranges
A_2, D_3, and D_4 are factors from Table 7.5

The point at which action should be taken should also be defined. While interpreting X-bar charts and R-charts, it may be decided that a red flag will be raised if one data point (the average of a subset) falls outside the control limits or if 5 to 8 consecutive subset averages fall on one side of the overall average. Periodically, it may be necessary to update the control limits from more recent data. This could be easily accomplished with spreadsheets or statistical software.

Example of an Average and Range Control Chart

This example illustrates the development of a statistical control chart, which tracks the percent passing the No. 50 sieve. A sieve analysis of the sand is conducted twice a day. Two tests *on different sand samples* are run on each day, and these two tests make up the subset of data with n = 2. The data are shown in Table 7.6. The first 30 subsets of data are used to calculate the control limits for the chart. Subsequent data are plotted using the control limits established from the historical data.

Control Limits for X-bar Chart (3-Sigma Limits):

Upper Control Limit

$$UCL_{\overline{X}} = \overline{\overline{X}} + (A_2 \times \overline{R}) = 22.2 + (1.88 \times 4.8) = 31.3$$

Lower Control Limit

$$LCL_{\overline{X}} = \overline{\overline{X}} - (A_2 \times \overline{R}) = 22.2 - (1.88 \times 4.8) = 13.1$$

TABLE 7.6
Control Chart for Percent Passing the No. 50 Sieve

Date	Test 1	Test 2	Average, $\bar{X}$	Range, R
May 1	22.0	19.2	20.6	2.8
May 2	18.8	20.9	19.9	2.1
May 3	30.7	16.6	23.7	14.1
May 4	21.0	22.5	21.8	1.5
May 5	15.7	24.7	20.2	9.0
May 6	13.8	16.0	14.9	2.2
May 7	19.2	16.7	18.0	2.5
May 8	24.0	24.1	24.1	0.1
May 9	28.6	23.5	26.1	5.1
May 10	23.3	30.5	26.9	7.2
May 11	22.0	30.2	26.1	8.2
May 12	18.1	21.9	20.0	3.8
May 13	25.8	27.6	26.7	1.8
May 14	24.6	21.9	23.3	2.7
May 15	24.1	19.4	21.8	4.7
May 16	30.8	22.7	26.8	8.1
May 17	18.2	9.2	13.7	9.0
May 18	20.7	18.2	19.5	2.5
May 19	16.8	24.1	20.5	7.3
May 20	23.3	22.9	23.1	0.4
May 21	28.8	20.1	24.5	8.7
May 22	20.5	15.4	18.0	5.1
May 23	19.1	15.9	17.5	3.2
May 24	25.8	23.3	24.6	2.5
May 25	28.1	31.0	29.6	2.9
May 26	29.2	27.1	28.2	2.1
May 27	20.2	10.4	15.3	9.8
May 28	19.6	16.7	18.2	2.9
May 29	19.8	27.5	23.7	7.7
May 30	27.6	32.8	30.2	5.2
Subgroups		30	$\bar{\bar{X}} = 22.2$	$\bar{R} = 4.8$

Control Limits for R-Chart:

Upper Control Limit

$$UCL_R = D_4 \times \bar{R} = 3.27 \times 4.8 = 15.8$$

Lower Control Limit

$$LCL_R = D_3 \times \bar{R} = 0 \times 4.8 = 0$$

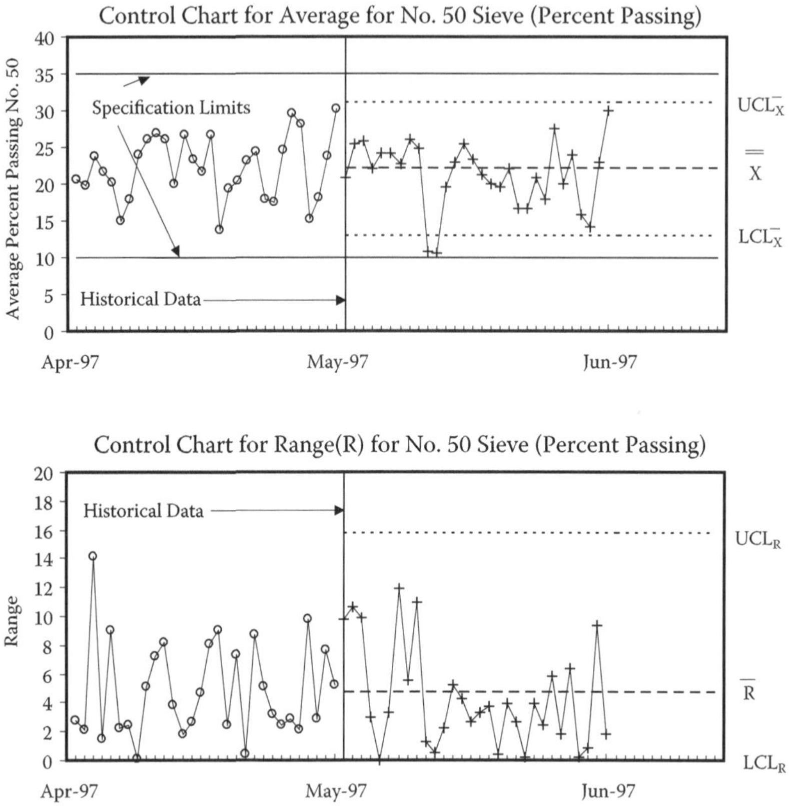

FIGURE 7.3 Example of statistical control chart.

Coefficients A_2, D_3, and D_4 are obtained from Table 7.5 for n = 2. The control charts are shown in Figure 7.3. Note that the average of each data subset is plotted on the X-bar chart. The specification limits for percent passing No. 50 is 10 to 35% and is also plotted on the X-bar chart.

Moving Average and Moving Range Charts

A control chart for moving average is identical to a control chart for averages. Typically, a moving average of 3 or 5 consecutive data points are plotted. The trend of a moving average chart may detect changes quicker than a control chart for averages. A moving average chart is not applicable for data arranged in subgroups such as in the previous example. However, the moving average chart is commonly used in tracking compressive strength test results (each being the average of two or more cylinders) from samples of different loads of concrete.

The moving range control chart can also be used to track within-batch variability. Other than this, all discussions provided for average and range control charts are also applicable for the moving average and moving range control charts. Even the method of calculating control limits is identical.

An example of an individual chart, moving average chart, and range chart is provided in the next section.

CUSUM Charts

Ken Day (2006) has been a major proponent of the use of CUSUM charts, which are commonly used in the United Kingdom and in Australia. A CUSUM chart works in the following manner: With any data set, an average is calculated. The difference between each individual value and the average is cumulatively summed and graphed against a timeline. As long there is no change in the slope of the chart (flat), the average value for that property plotted is constant. If the slope of the chart changes, it indicates a change in the average value, that is, a "step change." The extent of this step change can be estimated from the magnitude of the slope. A rising CUSUM chart (positive slope) indicates that there has been an increase in average value in that section. The point of intersection of the best straight line before and after the step change will pinpoint the time of occurrence of the step change. CUSUM charts are more effective than control charts in rapidly identifying step changes (Day 2006; Juran and De Feo 2010; Gibb and Garrison 2010; ACI 214R).

Example

Table 7.7 shows a list of randomly generated compressive strength test data. For sample numbers 16–30, a step change in the form of a reduction in average compressive strength of 750 psi was introduced as compared to sample number 1–15. Columns 3–5 in Table 7.7 show the methodology of CUSUM calculation as discussed in the previous paragraph. The CUSUM chart is plotted in Figure 7.4. The CUSUM chart clearly shows that the data varied in a random manner for the 15 samples, and the step change in the form a lower strength from sample 16 onward, and it continues till sample 30. The slope of the line is about –515 psi (3.5 MPa), which is lower than the actual step change of 750 psi.

The data in Table 7.7 is used for developing the control charts plotted as Figure 7.5. The control charts can be used to track the strength data collected on the job.

Figure 7.5a is an individual chart that plots the cylinder test results. The control limits plotted include

- the required average strength (f'_{cr}) (note that this is not the actual average of the results),
- the specified strength (f'_c), and
- (f'_c - 500) psi or (f'_c – 3.5) MPa.

The last limit is plotted to check whether an individual test falls below 2500 psi or 17.5 MPa, which is one of the ACI acceptance criteria for this example. This chart shows that none of the strength test results were below that strength test acceptance criterion. This chart shows random variation for the first 15 samples, and a step change seemed to have occurred from sample 16 onward. However, that is not clearly confirmed.

TABLE 7.7
Randomly Generated Compressive Strength Test Data Showing Step Change for Samples 16–30

Sample No.	Strength	Average Strength thus far	Difference from Average	CUSUM, psi	CUSUM, MPa
1	4376	4376	0	0	0
2	4080	(4376 + 4080)/2 = 4228	–148	–148	–1
3	3728	(4376 + 4080 + 3728)/3 = 4061	–333	–481	–3
4	3395	3895	–500	–981	–7
5	4270	3970	300	–681	–5
6	4979	4138	841	160	1
7	4080	4130	–50	110	1
8	3350	4032	–682	–572	–4
9	4819	4120	699	127	1
10	3363	4044	–681	–554	–4
11	3766	4019	–253	–806	–6
12	3981	4016	–35	–841	–6
13	3866	4004	–138	–979	–7
14	3378	3959	–581	–1560	–11
15	4775	4014	761	–799	–6
16	3626	3990	–364	–1162	–8
17	3330	3951	–621	–1783	–12
18	2978	3897	–919	–2702	–19
19	2645	3831	–1186	–3888	–27
20	3520	3815	–295	–4183	–29
21	4229	3835	394	–3789	–26
22	3330	3812	–482	–4271	–29
23	2600	3759	–1159	–5430	–37
24	4069	3772	297	–5133	–35
25	2613	3726	–1113	–6246	–43
26	3016	3699	–683	–6929	–48
27	3231	3681	–450	–7379	–51
28	3116	3661	–545	–7924	–55
29	2628	3625	–997	–8921	–62
30	4025	(4376+……..4025)/30 = 3639	386	–8535	–59

Figure 7.5b plots the moving average of three consecutive test results. The control limits plotted include

- the required average strength (f'_{cr}) and
- the specified strength (f'_c).

ACI acceptance criteria require that the moving average of three should not be less than the specified strength (f'_c). This control chart shows that 3 points were slightly

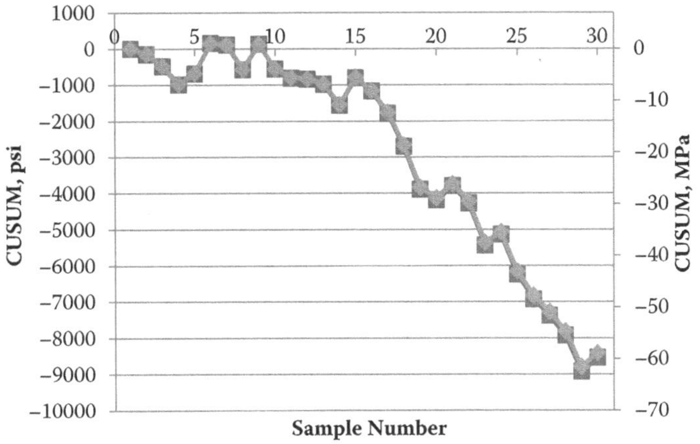

FIGURE 7.4 Example of CUSUM chart for strength test results in Table 7.7.

below that acceptance criterion. All 3 points occurred after sample 16. The step change after point 16 is clearer than in the individual control chart.

Figure 7.5c and d monitor the within-batch variation and a greater discussion on this topic, including the calculations shown in Table 7.8 is provided in the chapter on testing variability. Figure 7.5c plots the range of the two cylinders prepared from the same concrete sample expressed as a percent of the average strength. For this example, the control limit for acceptable range of two cylinders, plotted in the third chart, is 8%. Clearly, this level is exceeded much more often than 1 in 20, suggesting that cylinder fabrication and testing techniques need to be reviewed. Figure 7.5d plots the moving 10 test within-batch coefficient of variation (V_1) in percent. For this example, the testing quality is poor most of the time.

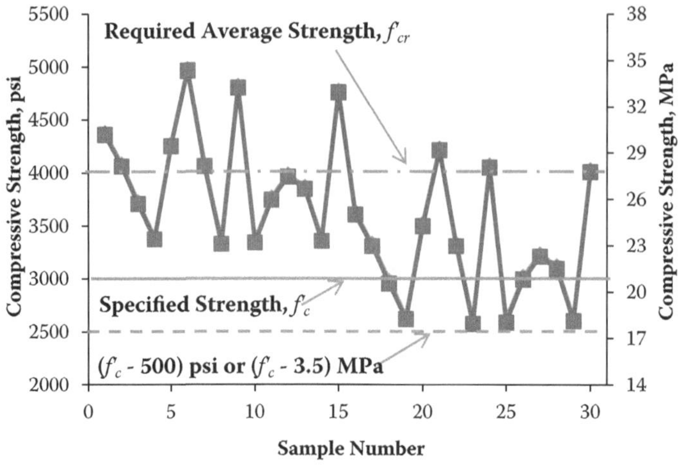

(a)

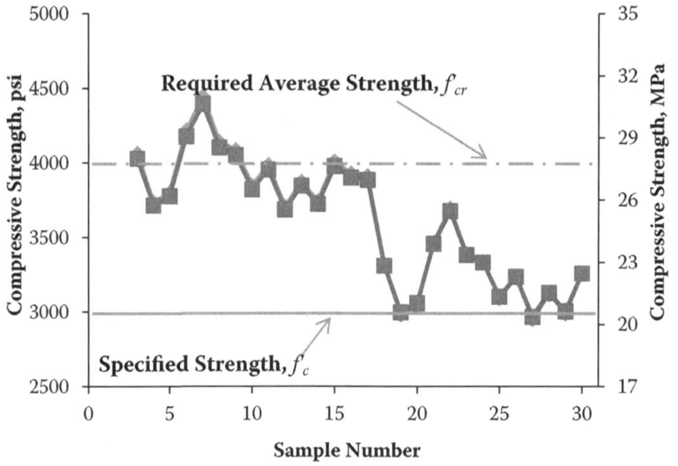

(b)

FIGURE 7.5 Example of control charts for strengths: (a) individual, (b) moving average of 3, (c) within-batch range calculated as percent of average strength, (d) within-batch coefficient of variation calculated in accordance with ACI 214R.

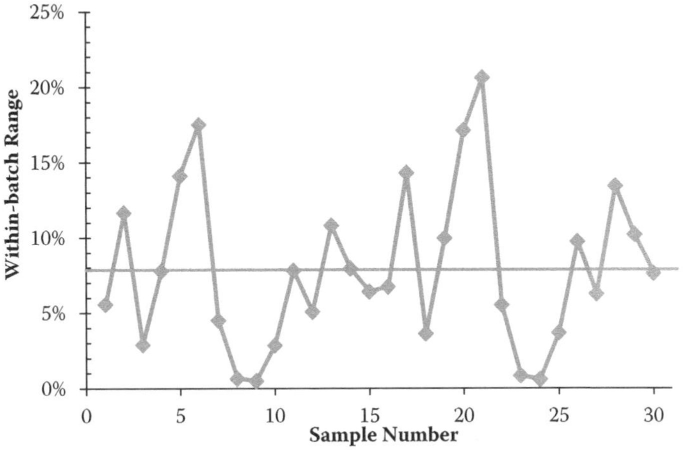

(c)

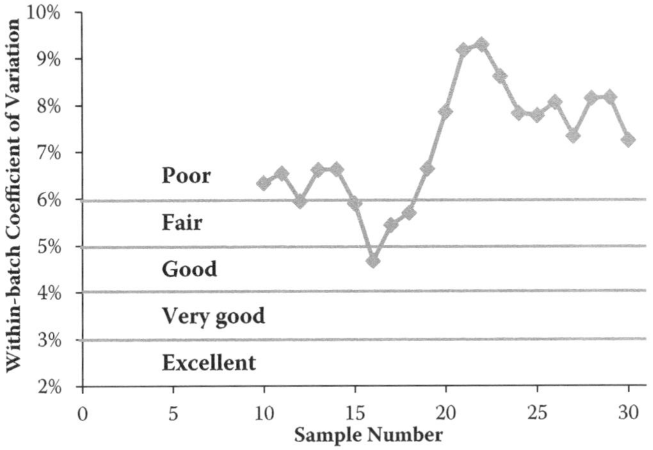

(d)

FIGURE 7.5 (continued) Example of control charts for strengths: (a) individual, (b) moving average of 3, (c) within-batch range calculated as percent of average strength, (d) within-batch coefficient of variation calculated in accordance with ACI 214R.

TABLE 7.8
Compressive Strength Test Data and Calculations for Control Chart Shown in Figure 7.5

Sample No.	Strength Cylinder 1	Strength Cylinder 2	Strength Test Results	Moving Average of 3 Strength Tests	Within-Batch Range	Within-Batch Range, %	Moving Average of 10 Strength Tests	Moving Average of 10 Ranges	Moving 10 Test V_1, %
1	4498	4254	4376		244	5.6%			
2	4318	3842	4080		476	11.7%			
3	3782	3674	3728	4061	108	2.9%			
4	3527	3263	3395	3734	264	7.8%			
5	4571	3969	4270	3798	602	14.1%			
6	4543	5415	4979	4215	872	17.5%			
7	3988	4172	4080	4443	184	4.5%			
8	3361	3339	3350	4136	22	0.7%			
9	4831	4807	4819	4083	24	0.5%			
10	3411	3315	3363	3844	96	2.9%	4044	289	$0.886 \times \frac{289}{4044} = 6.3\%$
11	3619	3913	3766	3983	294	7.8%	3983	294	6.5%
12	3880	4082	3981	3703	202	5.1%	3973	267	5.9%

13	4075	3657	3866	3871	418	10.8%	3987	298	6.6%
14	3244	3512	3378	3742	268	7.9%	3985	298	6.6%
15	4928	4622	4775	4006	306	6.4%	4036	269	5.9%
16	3748	3504	3626	3926	244	6.7%	3900	206	4.7%
17	3568	3092	3330	3910	476	14.3%	3825	235	5.4%
18	3032	2924	2978	3311	108	3.6%	3788	244	5.7%
19	2777	2513	2645	2984	264	10.0%	3571	268	6.6%
20	3821	3219	3520	3048	602	17.1%	3587	318	7.9%
21	3793	4665	4229	3465	872	20.6%	3633	376	9.2%
22	3238	3422	3330	3693	184	5.5%	3568	374	9.3%
23	2611	2589	2600	3386	22	0.8%	3441	335	8.6%
24	4081	4057	4069	3333	24	0.6%	3510	310	7.8%
25	2661	2565	2613	3094	96	3.7%	3294	289	7.8%
26	2869	3163	3016	3233	294	9.7%	3233	294	8.1%
27	3130	3332	3231	2953	202	6.3%	3223	267	7.3%
28	3325	2907	3116	3121	418	13.4%	3237	298	8.2%
29	2494	2762	2628	2992	268	10.2%	3235	298	8.2%
30	4178	3872	4025	3256	306	7.6%	3286	269	7.2%

8 Variation in Concrete Performance Due to Batching

This chapter discusses the benefits of controlling accuracy of batching of concrete material ingredients other than water. Effects of batching inaccuracies of water and ways to reduce it have been addressed earlier in Chapter 4 Mixing Water Control.

ASTM C94 SCALE ACCURACY AND ACCURACY OF PLANT BATCHING

ASTM C94-11 states that scales shall be considered accurate when at least one static load test within each quarter of the scale capacity can be shown to be within ±0.15% of the total capacity of the scale or 0.4% of the net applied load, whichever is greater. The NRMCA plant certification check list invokes these requirements through the range of use of scales. The certification requires companies to verify scale accuracy at least once every 6 months and to arrange for prompt recalibration and correction if noncompliance is indicated or if the plant is moved or other maintenance impacts the weighing systems. The accuracy of volumetric measuring devices in the NRMCA Check List for water and chemical admixtures is established by the required batching accuracy. NRMCA certification requires that volumetric measuring devices should be checked for accuracy at least once every 6 months.

ASTM C94 and ACI 117-06 state the tolerances for batching ingredients of concrete. The batching accuracy requirements in the NRMCA Plant Certification program are as required in ASTM C94 and stated in Table 8.1. For weighed ingredients, accuracy of batching is determined by comparison between the desired weight and the actual scale reading; for volumetric measurement of water and admixtures, accuracy is determined by checking the discharged quantity relative to the target, either by mass on a scale or by volume in an accurately calibrated container. Volumetric measurement is commonly used for water and typically used for chemical admixtures. It is realized that in any single batch, at least one of the ingredients may be out of tolerance, as discussed later. For the NRMCA plant certification, inspectors are advised to review several batch records and to determine that the plant complies with the batching accuracy by reviewing the average of any 10 consecutive batches of concrete.

TABLE 8.1
Batching Accuracy Requirements per ASTM C94 and NRMCA Plant Certification

Summary of Requirements for Accuracy of Batching

Material	Cementitious Materials	Aggregates: Individual Batchers	Aggregates: Cumulative Batchers	Water, Vol. or Wt.	Admixtures
Basic tolerance	±1% of desired weight	±2% of desired weight	±1% of desired weight	±1.5% of desired value or	±3.0% of desired value or
	Intermediate and cumulative		Intermediate and cumulative	1.0 gal (10 lb)	Minimum dosage rate per 100 lb cement
Small batch tolerance	−0 to +4% of desired weight	±0.3% of scale capacity for loads below 15% of scale capacity	±0.3% of scale capacity for loads below 30% of scale capacity	Whichever is greater	Whichever is greater

Source: Adapted from NRMCA Quality Control Manual—Section 3, "Checklist for Certification of Ready Mixed Concrete Production Facilities," Eleventh Revision, May 2011, NRMCA, www.nrmca.org.

TWO ISSUES WITH BATCHING

The above requirements are essential first steps for ensuring batching accuracy. Concrete producers should continuously review batch records to ensure that batch weights of all the ingredients of concrete are within the ASTM C94 batching accuracy requirements. Concrete plants may have to be tuned, continually monitored, and adjusted when necessary. If this is not practiced, two possible errors may occur:

1. Over-batching materials means giving material away and increased material cost per cubic yard produced. Under-batching results in under-yield causing customer complaints.
2. Batch weights that are highly variable can cause significant variations in yield, strength, and other performance characteristics of concrete. It results in poor inventory control of ingredient materials at the plant.

OVER-BATCHING

Figure 8.1 (Bain and Obla 2007) shows the cement over-batch in dollars from a fairly new concrete plant producing approximately 200 yd³ (150 m³) per day. It is clear that for the first month (June 1 to 30) the plant was on an average over-batching about 10 lbs/yd³ (6 kg/m³) of cement. Between June 30 and September 30, the over-batching continues upward in a series of ever-shortening steps. This was due to attempts to

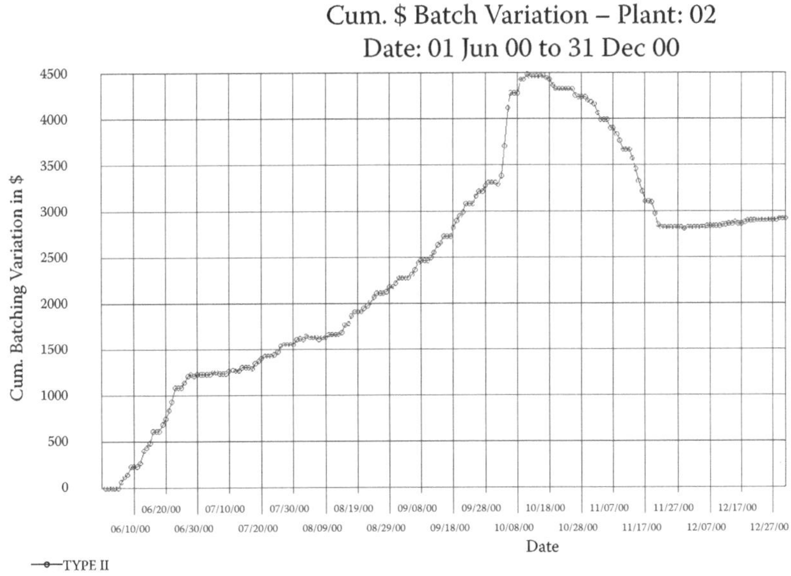

FIGURE 8.1 Significant cement over-batching prior to installation of improved QC system (1 plant producing about 200 yd³ [150 m³] per day working 5 days/week). (Reprinted with permission from Bain, D., and Obla, K. H., "Concrete Quality Control—The Untapped Profit Center," *Concrete InFocus*, Fall 2007, Vol. 6, No. 3, NRMCA, pp. 63–69.)

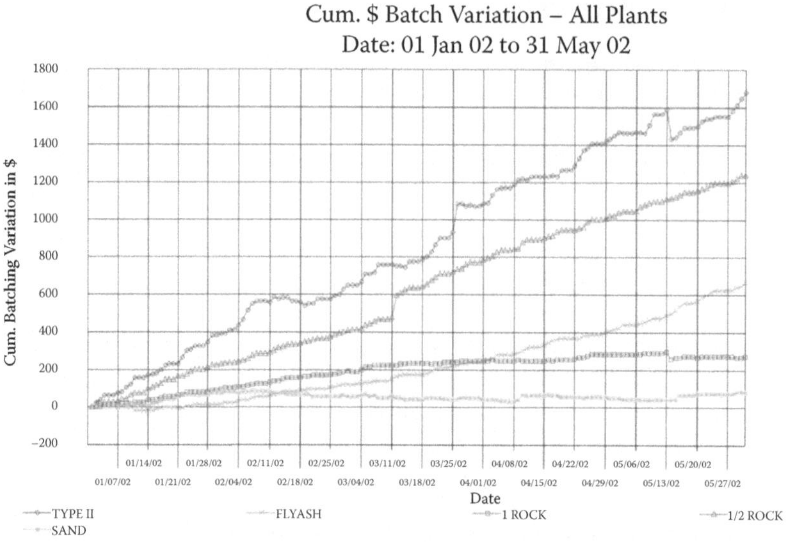

FIGURE 8.2 Significant reduction in material over-batching after installation of improved QC system (5 plants producing total of about 300,000 yd^3 [230,000 m^3] for that period). (Reprinted with permission from Bain, D., and Obla, K. H., "Concrete Quality Control—The Untapped Profit Center," *Concrete InFocus*, Fall 2007, Vol. 6, No. 3, NRMCA, pp. 63–69.)

tune the plant using the batch computer as the plant continued to become more and more mechanically unsound. It can be easily seen at what point the plant fails. It is also apparent that once the plant was repaired to a proper mechanical condition, that the batch computer had in fact been tuned to an under-batch condition to correct a deteriorating mechanical condition that had been causing the plant to over-batch. Once the plant was repaired and the computer properly tuned, more accurate batching was possible as reflected by the almost flat line starting about November 20.

Figure 8.2 (Bain and Obla 2007) represents what can be achieved in batching accuracy when attention is given to all materials. This figure represents five plants producing a total of almost 3000 yd^3 (2300 m^3) per day. Note that this figure represents 5 months of production totaling over 300,000 yd^3 (230,000 m^3). Total material over-batch was reduced to an average of \$0.013/yd^3 (\$0.017/m^3) for the 5-month period. This type of performance requires constant monitoring and constant preventative maintenance.

Another producer (Crammatte 1998) reports similar reductions in over-batches. Figure 8.3 (upper plot) is a timeline graph (over 2 years) of sample averages of cement batch weights of three consecutive batches of concrete. Each data point was the average of cement batch weights for three consecutive concrete batches for the same mixture. At least two such points were collected per day, so the whole plot consists of 663 data points. At the beginning, cement was being over-batched on average about 5 lb/yd^3 (3 kg/m^3) over the target weight.

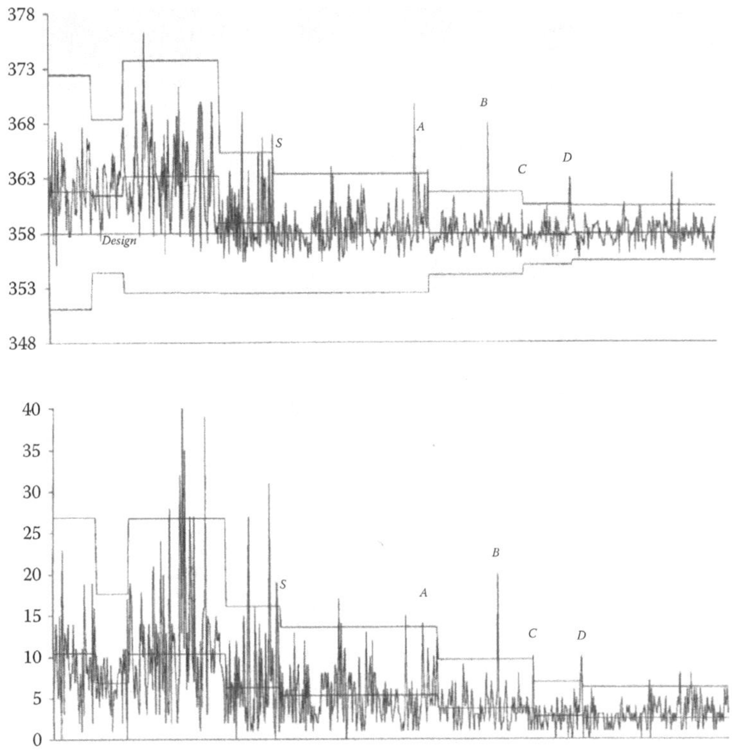

FIGURE 8.3 Control charts for the cement batching process over 2 years at a ready mixed concrete plant. (Reprinted with permission from Crammatte 1998.) Average cement weight in lb/yd^3 of three consecutive batches of concrete (upper) and corresponding range in lb/yd^3 (lower). Two data points plotted each day of production.

Variation of Batch Weights and Its Effects

The second important issue with batching is the variation in batch weights. In 2006, a survey of batching accuracy was conducted (Daniel 2006) while developing the NRMCA Plant Inspectors Guide for the purpose of evaluating how inspectors can check batching accuracy during a plant inspection. Data were received from 31 concrete plants with batch records from 10 consecutive batches for each of 2 days. This produced 620 batch records. Actual batch weights were compared with target weights for all of the concrete ingredients—cement, fly ash, slag cement, coarse aggregates, fine aggregates, air entraining admixtures, and two other chemical admixtures (if used). Out of 620 batches, 565 batches were 3 yd^3 (2.3 m^3) or higher, and the following analysis is restricted to those since batches less than 3 yd^3 (2.3 m^3) are more prone to batching inaccuracies. Table 8.2 provides the analysis of the out-of-tolerance batches for just the cementitious weight. It was found that 83% of all

TABLE 8.2
Analysis of Cement Batching from a NRMCA Batching Accuracy Survey

Tolerance Limits, %	Percent within Tolerance, %
±1%	83
±2%	95
±5%	98
±10%	99

Source: Adapted from Daniel, G., "Batch Record Data Compilation and Summary," submitted to the RMC Research and Education Foundation, 2006.
Note: C94 cement batching tolerance is ±1%.

batches were within the C94 tolerance of ±1% and 95% were within ±2%. When 10 consecutive batch records were averaged, 98.5% of the plants met the batching accuracy requirements of ASTM C94. Another aspect of this review is that most batches were out of tolerance on at least one ingredient.

Cementitious Weight Variation and Its Effect on Strength Variation

For a given concrete mixture, variations in cementitious batch weights can be expected to lead to strength variations. As discussed earlier, a commonly used rule of thumb is that 1 lb of cementitious material equates a compressive strength between 8 and 12 psi (or 1 kg of cementitious material equates a compressive strength between 0.12 to 0.18 MPa). This means that a ±1% variation in cementitious batch weight will result in a ±1% compressive strength variation. If the cementitious batch weight varies by ±1% of the target value 95% of the time, the effect is that the resulting compressive strength variation due to just the cementitious batch weight variation should be within ±1% of the average strength 95% of the time. The variation in mixing water content will have an exactly similar but inverse relationship with compressive strength variation and has been discussed in an earlier chapter.

For the purpose of this analysis, it is also assumed that the cementitious batch weight and the resulting compressive strength of the concrete are normally distributed. If 95% of the time it is within tolerance, then it follows that 97.5% of the time the cementitious batch weight is more than the lower tolerance limit, and as a result, 97.5% of the time the compressive strength of the concrete should be greater than the strength corresponding to the cementitious batch weight at the lower tolerance limit.

So, for the 97.5% probability, it follows that:

$$\bar{X} - 1.96S = (1-y) \times \bar{X}$$

where $\bar{X}$ is the average compressive strength; y = variation in cementitious batch weight (0.01 for 1%; 0.02 for 2% etc.); S = standard deviation of resulting compressive strength test results.

TABLE 8.3
Cementitious or Mixing-Water Content Variation and Its Effect on Concrete Strength Variation and Materials Cost

Cementitious or Mixing-Water Batch Weight Variation,%	Calculated Concrete Strength Standard Deviation, psi[a] (MPa)	Required Average Strength for $f_c' =$ 5000 psi (34.5 MPa)	Cost Savings Due to Better Batching Control, \$/yd^3 (\$/m^3)
±1%	26 (0.18)	5034 (34.7)	1.67 (2.18)
±2%	51 (0.35)	5070 (35.0)	1.49 (1.95)
±5%	128 (0.88)	5177 (35.7)	0.95 (1.24)
±10%	255 (1.76)	5367 (37.0)	0

[a] The standard deviations of the compressive strengths are calculated by assuming that the cementitious or mixing-water batch weights are within the noted variation 95% of the time, and the batch weights and resulting concrete compressive strengths are normally distributed.

Simplifying, we get

$$S = \frac{y}{1.96}\bar{X}$$

If $\bar{X}$ = 5000 psi (34.5 MPa), S can be calculated as given in Table 8.3 for the various cementitious batch weight variations. Table 8.3 shows that the resulting strength standard deviation varies linearly with the cementitious or mixing-water batch weight variation. If the cementitious batch weight varies by ±1% of the target value 95% of the time, the resulting strength standard deviation will be 26 psi (0.18 MPa). If the cementitious batch weight varies by ±2% of the target value 95% of the time (findings from the survey reported in Table 8.2), the resulting strength standard deviation will be 51 psi (0.35 MPa). Table 8.3 also shows the required average strength for various cementitious or mixing-water batch weight variations for a specified strength of 5000 psi (34.5 MPa). It is clearly evident that reducing the cementitious or mixing-water batch weight variation from ±10% to ±5% will result in 190 psi (1.31 MPa) lower required average strengths. Assuming that every 200 psi (1.38 MPa) increase in compressive strength requires an increase in materials cost of \$1/yd^3 (\$1.31/m^3), a 190 psi (1.31 MPa) reduction in required average compressive strength will result in a material cost savings of \$0.95/yd^3 (\$1.24/m^3). Reducing the batching variation further to ±1% will lead to even more material cost savings.

If a concrete producer is reviewing batch records, a suggested target for improvement would be to ensure that the cementitious batch weights should be within ±1% in 95% of the batches. As a process of continuous improvement, once the 95% target level has been attained, producers can aim for 99%. One company (Crammatte 1998) has reported attaining a batching accuracy of 99.9999% or only 1 in a million batching failure rate and has quantified significant material cost savings as should be expected!

While cement is usually the primary focus of any plant analysis, all materials are subject to variations in batching and all have an effect on the quality of the concrete. Batching inaccuracy can be cumulative. A 2% reduction in cementitious batch weight, in addition to a 2% increase in aggregate batch weights in the same batch, can lead to a 4% reduction in cementitious weight when concrete is adjusted for yield, thus resulting in significant concrete performance variation. Also admixture dosage variations can lead to significant changes in air content and mixing water content.

HOW CAN A COMPANY IMPROVE BATCHING ACCURACY?

Once a target has been established, there are different ways companies can go about to improve batching accuracy. Bain and Obla (2007) report the use of a real-time error monitoring system with great success. Control systems available to the industry allow the notification of responsible company personnel, in the form of an email, whenever the set batch tolerances are exceeded. Parameters for these alerts can be set by recipient, region, plant, material, and magnitude of the error. These alerts arrive in the hand of the intended recipient (quality personnel) in real time so that a decision can be made whether to correct the error, to prevent that particular batch of concrete from being delivered to a project, to divert the load to another customer, or to discard the load. Real-time reporting of batching errors is a good way to monitor the changing mechanical condition of a plant as well.

Crammatte (1998) has had significant success in improving batching accuracy using statistical process control charts. Cramatte used the control charts for average and range of ingredient batch weights of three consecutive batches of concrete. In the process described, the plant operator was asked to plot two points each of three consecutive batches a day from the batch weight data being reported by the batch control computer. A top-selling concrete mix was picked for the evaluation. The first two batches of the day were not used, as plant equipment had not yet warmed up. When the data collection was initiated, quick changes in plant operations were avoided. Changes were made methodically and involved resetting constants in the batch computer without any hardware changes. Figure 8.3 shows control charts for the cement batching process. The point "S" on the charts in Figure 8.3 is the point where this process was brought into statistical control most of the time. The points under A, B, C, and D were caused by the batch control computer changing its own constants, reacting to assignable cause variation as though it were a shift in the process average, and adjusting where no adjustment was warranted. Initially, there were differences as high as 40 lb/yd^3 (23.7 kg/m^3) between batches that could have led to significant variation in concrete performance, but over time, this was consistently reduced to about 5 lb/yd^3 (3 kg/m^3), which is about an 80% reduction in the range of the cementitious batch weight. Similar control charts were developed for all the ingredient materials. Figure 8.4 shows the control charts for the fine natural dune sand batching process. Changes to batching, such as tolerances, jog timing, jog duration, and time in air fall, were made methodically by the plant operator. The overall result was an 80% reduction in the average range. In the beginning there was difference as high as 80 lb/yd^3 (50 kg/m^3) between batches, significant enough to cause quite a bit of variation in yield, strength, and so forth. In the lower chart, the four "out of control" points under the letter A are caused by one batch in the sample going way over

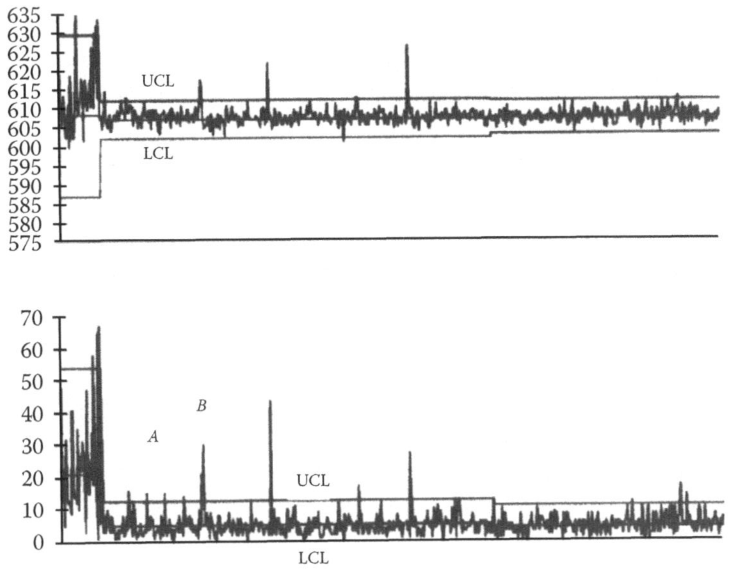

FIGURE 8.4 Control charts for the fine aggregate batching process over 2 years at a ready mixed concrete plant. (Reprinted with permission from Crammatte 1998.) Average fine aggregate weight in lb/yd^3 of three consecutive batches of concrete (upper) and corresponding range in lb/yd^3 (lower). Two data points plotted each day of production.

target batch weight just after the plant ran out of material on the previous batch. The "out of control" point under the letter B was from a modification tried. Overall, for all the ingredient materials, the average range was reduced by about 60%. For ingredients batched at very low weights (<100 lb/yd^3 (60 kg/m^3), it was harder to achieve large reductions. By methodically following the process and reducing the variations, the producer was able to operate at a variation less than half of the C94 tolerance for cement and aggregate batch weights, thus ensuring that tolerance limits will be exceeded less than one in a million batches! The producer reports success in implementing statistical process control in aggregate production and paperwork as well.

The producer states that an environment of trust that is free of fear is essential for statistical process control to succeed. Initially, rapid progress can be made by making simple changes, and it is important not to blame someone or some group because the low-hanging fruit will not be accessed. The producer states that a plant operator achieved most of the improvements shown in Figures 8.3 and 8.4, and has become an advocate of continuous improvement and pride in workmanship, traits that will benefit the company.

Another way of doing batching accuracy is by the cumulative end-of-day, -week, or -month method. If, at the end of some time period, the total amount of material used in the concrete more or less equals the amount that should have been used, then all is deemed to be well. This is a business-side argument. Unfortunately, this "inventory-based" method is not useful enough to adequately establish the true performance of a concrete plant or to predict the performance of individual concrete batches. There needs to be a quality side of the argument, too. A cumulative ending number only tells a small part of the story; it is necessary to determine the path to that number.

YIELD MEASUREMENTS—A TOOL TO IMPROVE BATCHING ACCURACY

Yield is the actual volume of concrete produced. The basis of sale for ready mixed concrete is by volume (yield), and its estimation is stated in ASTM C94. Yield can be calculated by dividing the weight of all materials batched by the measured density of concrete. Relative yield is the ratio of actual volume of concrete produced, compared to the volume the batch was designed to produce. It is suggested that the concrete producer make frequent yield measurements of their top two selling mixtures at each concrete plant. As long as the relative yield is within ±2% (i.e., between 0.98 and 1.02), the yield is probably within expected variation. A higher variation can suggest poor batching accuracy among other things. A greater discussion of concrete yield can be found elsewhere (NRMCA TIP 8 [2012], NRMCA CIP 8 [2000]). Even if the average (>10 measurements) relative yield is 1.01, it indicates that on average there is material over-batching by 1% (1% higher material costs) and lost sales of 1% (1% lost revenue).

SUMMARY

Improving batching accuracy can help reduce material over-batching and thus reduce material costs per cubic yard produced. It can also help reduce material under-batching and thus under-yield that could result in poor customer relations. Improving batching accuracy can help attain more consistent performance for fresh and hardened concrete properties, including variation of compressive strength of concrete. The intangible benefits of improving batching accuracy are also considerable. There will be fewer loads batched in error, and a reduction in rejected loads. Because of the constant tracking of material batching, it is possible to quickly detect plants that have just had a breakdown or where one is about to occur. This helps reduce plant downtime and maintenance.

9 Variation in Concrete Performance Due to Manufacturing

This chapter discusses control of variation associated with manufacturing specifically mixing temperature and delivery time. Batching accuracy, which is a part of manufacturing variation, was addressed in the previous chapter.

ASTM C94 REQUIREMENTS FOR UNIFORMITY OF CONCRETE

Concrete needs to be batched and mixed to a homogeneous mixture to reduce performance variation within a batch. Consistency in manufacturing concrete will also reduce variations between batches. ASTM C94 has requirements for uniformity of concrete produced by truck-mixing, shrink-mixing, or central-mixing. Two separate samples from a single batch of concrete, each consisting of approximately 2 ft^3, are taken after discharge of approximately 15% and 85% of the load. A series of concrete tests—slump, air content, compressive strength, coarse aggregate content, concrete and mortar densities on an air-free basis—are conducted. Limits are set for the range of results for each test on the two samples as listed in Table 9.1 (reproduced from ASTM C94) to meet the requirements for uniformly mixed concrete. ASTM C94 requires that uniformly mixed concrete should be achieved in a truck mixer with 70 to 100 revolutions at the mixing speed designated by the manufacturer. For plant mixers, it establishes a minimum mixing time that can be shorter when mixer qualification tests are performed. If the uniformly mixed concrete is not achieved within the minimum mixing requirements, after all ingredients including water are in the mixer, then the mixer shall not be used until the condition is corrected. This is discussed in the next section. Mixing uniformity tests are performed on mixers of a unique design by manufacturers that provide plant and truck mixers in accordance with the standards of the Concrete Plant Manufacturers Bureau and the Truck Mixer Manufacturers Bureau, respectively. Concrete mixtures with low and higher slump levels are tested. When satisfactory performance is shown in one mixer, the performance of mixers from the same manufacturer, similar drum size, and blade design can be assumed, provided the mixer is at a similar or better overall condition (Daniel and Lobo 2005). The primary factors relative to the condition of mixers in operation that adversely impact efficiency in mixing are the blade wear and buildup of hardened concrete or mortar inside the drum. These two items are part of the inspection process for mixers in the NRMCA plant certification program. Additionally, plant mixers need to be evaluated for mixing uniformity by testing

TABLE 9.1
Requirements for Uniformity of Concrete (ASTM C94)

Test	Requirement, Expressed as Maximum Permissible Difference in Results of Tests of Samples Taken from Two Locations in the Concrete Batch
Mass per cubic foot (mass per cubic meter) calculated to an air-free basis, lb/ft^3(kg/m^3)	1.0 (16)
Air content, volume of concrete	1.0
Slump:	
If average slump is 4 in. (100 mm) or less, in. (mm)	1.0 (25)
If average slump is 4 to 6 in. (100 to 150 mm), in. (mm)	1.5 (40)
Coarse aggregate content, portion by mass of each sample retained on No. 4.75 mm sieve	6.0
Mass per unit volume of air-free mortar based on average for all comparative samples tested (%)	1.6
Average compressive strength at 7 days for each sample,[a] based on average strength of all comparative test specimens (%)	7.5[b]

Source: Reprinted with permission from ASTM C94, "Standard Specification for Ready Mixed Concrete," Volume 4.02, ASTM International, West Conshohocken, PA, 2009, www.astm.org.

[a] Not less than 3 cylinders will be molded and tested from each of the samples.

[b] Approval of the mixer shall be tentative, pending results of the 7-day compressive strength tests.

slump and coarse aggregate content for plants that want to be qualified as central-mixing plants (NRMCA Quality Control Manual—Section 3, 2011).

IMPROVING UNIFORMITY OF CONCRETE PRODUCED IN TRUCK MIXER

Generally, central-mixed concrete produced in plant mixers will not have variations in the equipment and operator influence associated with truck-mixed concrete. NRMCA, in consultation with truck and plant manufacturers, investigated a wide number of factors that could have an effect on the uniformity of concrete produced in a truck mixer. A series of 670 concrete batches were tested between 1969 and 1972 (Bloem and Gaynor 1970; Gaynor and Mullarky 1975). The following were determined to be the main factors, besides the design and condition of the mixer, that affect the homogeneity of a given batch of truck-mixed concrete: batching sequence, mixing speed, and mixing revolutions.

Batching Sequence

The following method is suggested to help improve uniformity of concrete as well as to avoid cement balls and head packing (Gaynor 1996):

1. Load about 4000 lbs (2000 kg) of coarse aggregate. This will avoid cement balls and head packs. Cement balls are round lumps of cement and fine and coarse aggregates, typically about the size of a baseball. A head pack is a combination of fine aggregate and cement packed against the head of the mixing drum. Batching sand or cement first causes head packs that cannot be mixed out and sometimes breaks lose after about half the load has been discharged causing variations in concrete uniformity.
2. Add three quarters of the batch water, while ribboning the remainder of the aggregate and cement. Typically, the batch water starts early, stops during cement batching, and ends well before the last of the aggregate.
3. Add the last quarter of the batch water to wash all materials in the charge hopper and discharge end into the mixer.

These water addition proportions may be used as a starting point, and if the ASTM C94 concrete uniformity requirements are not attained, then the proportions of the water may be adjusted. However, if too much water is added at the end, concrete at the discharge end will be wetter as the water will not move toward the head of the drum because the bulk of the concrete has not reached an adequate slump and the necessary flow pattern will not develop.

Mixing Revolutions

ASTM C94 requires that the uniformity requirements of concrete mixed in a truck mixer be met with 70 to 100 revolutions. Increasing the revolutions does not always improve uniformity, especially if a proper batching sequence is not used. ASTM C94 requires that additional revolutions of the mixer (during transit and waiting) beyond the number found to produce the required uniformity of concrete shall be at a manufacturer-designated agitating speed, which is usually between 2 and 6 revolutions per minute (rpm). This is because additional revolutions at mixing speed can cause heat buildup in the concrete and a reduction in both slump and air content. A few turns at mixing speed before discharge can enhance uniformity after a long or bumpy ride from the concrete plant to discharge (Daniel and Lobo 2005).

Mixing Speed

Mixing speed is commonly in the range of 6–18 rpm. Research work conducted at NRMCA (Bloem and Gaynor 1970; Gaynor and Mullarky 1975) demonstrated that for a fixed number of total drum revolutions, varying the drum speed in the range of 4 rpm to 12 rpm did not significantly affect the uniformity. However, uniformity improved in the range from 12 rpm to 22 rpm.

In a ready mixed concrete truck, the use of spiral blades moves the concrete first down toward the head end of the drum then back up the central axis toward the discharge end (Gaynor 1996). This causes a folding action that blends the ingredients together. If the mixing speed is not optimal, then this flow pattern is not created inside the truck. This desirable flow pattern occurs only at reasonably high mixing speeds (typically 18 rpm to 22 rpm but as low as 12 rpm to 15 rpm with some mixers). To observe whether the necessary mixing action and material flow is being obtained, one can look into the discharge end of the drum using a flashlight while the drum is revolving at mixing speed. Goggles and appropriate safety precautions should be followed if this is done. At a low drum speed, the concrete surface will be level with the ground. Increase drum speed in 1 rpm or 2 rpm increments until the concrete surface changes from level to almost perpendicular to the drum axis. If this mixing action is consistently produced, concrete can be mixed homogeneously in as little as 40 to 50 revolutions.

Concrete is frequently retempered at the job site. During the NRMCA research (Gaynor and Mullarky 1975), it was found that when 2.5 gallons/yd^3 of water was added followed by mixing at 10 rpm for 20 revolutions, the uniformity of concrete was only fair; however, when mixing was increased to 16 rpm, as few as 5 revolutions were sufficient to produce acceptable uniformity. It is clear that high mixing speed can help attain uniformity of concrete more efficiently and in a much shorter time period. However, if the mixing speed continues to increase beyond 22 rpm due to increasing centrifugal forces, the optimal flow pattern may not be achieved and mixing and concrete uniformity starts to deteriorate (Bloem and Gaynor 1970). The mixing speed on current mixers might be limited to about 20 rpm to reduce the weight of trucks and maximize load size.

What Can a Company Do to Improve Uniformity of Concrete Produced in a Truck Mixer?

Companies should first realize that that to produce quality concrete with a low variation in performance the concrete needs to be homogeneously mixed. They can achieve that with the following steps:

1. For every truck, do an annual check for blade wear and concrete buildup. These two items are part of the inspection process for truck mixers in the NRMCA plant certification program. Blade wear should be checked at the point of maximum drum diameter nearest to the drum head. When the height of the blade at this point, measured from the drum shell, is less than 90% of the original radial height (dimension "X" in sketch of applicable blade type in Figure 9.1), the blade is considered excessively worn. The manufacturer of the mixer will furnish original blade dimensions on request. Manufacturers may have alternative recommendation for excessive blade wear for their equipment. The interior of the mixing drum must have more than a majority of the drum wall clean. The acceptable buildup of hardened concrete should typically be less than about 1500 lb (680 kg) or 3/8 yd^3 (0.3 m^3). Considering concrete to weigh about 150 lb/ft^3 (2400 kg/m^3), if the mixing drum belly has 4 lineal feet (1.2 m) of its complete circumference coated with 1 in. (25 mm) thickness of concrete, the weight is about 1500 lb (680 kg). If the head plus 2 ft (0.6 m) of the wall is coated 1¾ inch (45 mm) deep,

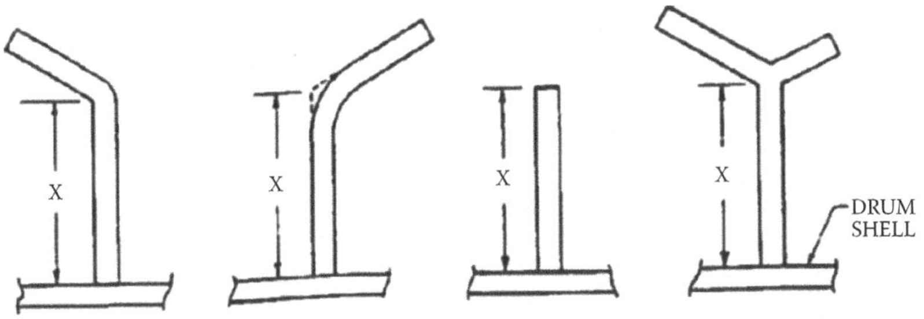

FIGURE 9.1 Mixer blade types. (Reprinted with permission from NRMCA Quality Control Manual—Section 3, "Checklist for Certification of Ready Mixed Concrete Production Facilities," Eleventh Revision, May 2011, NRMCA, www.nrmca.org..)

the total accumulation is 1500 lb (680 kg). Typically, the largest buildup on the blades will be on the mixing face (side facing the drum head) and may be difficult to observe. Blades at the discharge end and chutes should also have a minimum amount of buildup so as to permit ease of discharge without segregation.

2. The company may choose to perform a periodic mixing uniformity evaluation on mixers. This can be used to establish a critical condition of the mixer for both blade wear and buildup that impacts mixing efficiency. This evaluation should include slump, air content, and compressive strength, and the range of results should be compared to the limits established in ASTM C94. The complete set of mixing uniformity tests is rarely necessary because these are related. Since a large portion of the permitted difference can be associated with testing variation, a skilled technician who has an ACI Field Grade I certification should perform these tests. Alternatively, a visual observation of slump during discharge at the job site, possibly accompanied by occasional testing, is adequate to evaluate basic mixing uniformity. If problems with achieving mixing uniformity cannot be attributed to the condition of the truck mixer, it is likely to be due to a poor batching sequence, which may have to be rectified as discussed earlier.
3. Mixing speed should be selected so that the desired flow pattern is created inside the truck. Flow patterns can be studied at the same time random concrete uniformity evaluations are being conducted as discussed earlier.

VARIATION IN CONCRETE PERFORMANCE DUE TO TEMPERATURE

Temperature is another key variable that can lead to variation in concrete performance. Depending on ambient conditions concrete temperatures can vary from 40°F to 95°F during the course of a year and can still be used. As concrete temperature increases it is well known that mixing-water demand increases, slump loss increases, air content decreases, setting time decreases, early-age strength increases, and later age strength decreases. Later age strengths are not a concern as long as the cylinders are cured between 60°F and 80°F, that is, under standard curing conditions in accordance with ASTM C31 immediately after they are made.

Effect of Temperature on Setting Time

Setting time is a concern primarily for finishing slabs and contractors desire consistent setting times. As a rule of thumb, for every 20°F (11°C) increase in concrete temperature, setting time as measured by ASTM C403 decreases by about half. Contractors expect the concrete to have gained sufficient stiffness to commence finishing operations within x hours after the materials are batched at the plant. The value "x" may be discussed at a preconstruction meeting between the producer and contractor and may typically be about 2–3 hours after the truck arrives at the job site. Several researchers have concluded that finishing operations occur much before initial set (Bury et al. 1994; Abel and Hover 2000; Suprenant and Malisch 1998), which is the time it takes for the penetration resistance to reach 500 psi (3.5 MPa). Contractors have suggested (Malisch 2010) that an ASTM C403 setting time at a penetration resistance of 50 psi (0.35 MPa) is a better indicator for timing finishing operations than 500 psi (3.5 MPa) because by then all finishing operations should be completed. However, since the C403 testing variation at low penetration resistances are unknown, it is suggested that producers target ASTM C403 setting time at 500 psi (3.5 MPa) of (x ± 1) h. Relative setting time of concrete can be estimated in an automated manner using thermal measurements of concrete (Cost 2009).

Producers have some control of the concrete temperature at the concrete plant and can reasonably estimate the temperature change during transit to the job site. However, after the concrete is placed, the setting characteristics of the concrete can be affected by the subgrade temperature, ambient temperature, and radiant heat due to exposure to sunlight. So it can become hard to predict the concrete temperature once it is placed to attain consistent setting times in the field. Producers have two options:

1. They can estimate concrete setting times based on the concrete temperature at the plant.
2. They can estimate concrete setting times based on the average of the concrete temperature at the plant and ambient temperature expected after time x.

Producers can start with the first option and attempt the second option to see whether it improves the consistency of field setting times further.

Table 9.2 shows how a producer can implement this approach. Let us say for a particular mixture the measured ASTM C403 setting time at 500 psi (3.5 MPa) at 70°F (21°C) was 4 h and the target setting time is 4 ± 1 h. The second column shows how the setting time can change based on the rule of thumb. It is clear that even a 5°F (3°C) change in concrete temperature (this can occur during the course of a day) can make it difficult to meet the target setting times. For varying concrete temperatures, producers should develop beforehand a table of ASTM C494 admixture type (accelerator and retarder), dosage level, and mixture modification (if any) that would

TABLE 9.2
Control of Setting Time and Water Demand of Concrete as a Function of Temperature

Concrete Temperature, °F (°C)	ASTM C403 Setting Time,[a] h	Suggested Admixture Type and Dosages for Consistent Set Times, oz/cwt (mL/100 kg)	Water Demand,[b] lb/yd³ (kg/m³)	Suggested Admixture Type and Dosages for Consistent Water Content, oz/cwt (mL/100 kg)
40 (4.4)	12	Type C @ 45 (2930) + Lower SCM	275 (163)	Reduce cementitious content by 2%
50 (10.0)	8	Type C @ 30 (1956)	280 (166)	No admixture
60 (15.6)	6	Type C @ 10 (652)	285 (169)	No admixture
70 (21.1)	4	None	290 (172)	Type A @ 2 (130)
80 (26.7)	3	Type B @ 2 (130)	295 (175)	Type A @ 4 (261)

Note: Values in this table should not be taken as specific recommendations as they may not be applicable to local materials.
Admixture dosage is reported as ounces per 100 lb of cementitious or mL per 100 kg of cementitious.

[a] Based on rule of thumb that for every 20°F (11.1°C) increase in concrete temperature setting time decreases by half.

[b] Based on past NRMCA research (Gaynor et al. 1985) that every 10°F (5.6°C) increase in concrete temperature increases the mixing-water demand increases by about 5 lbs/yd³ (3 kg/m³).

help to bring the concrete mixture to the target setting times. A possible schematic is shown in Column 3 in Table 9.2. Generally, it is more economical to retard the concrete compared to accelerating it using nonchloride accelerating admixtures. It is possible that for the same concrete mixture at cooler temperatures (say, 55°F [13°C] and below), the contractor may accept a slightly higher target setting time of 6 ± 1 h instead of 4 ± 1 h. ACI 301 states that during cold weather concrete for slabs that are less than 12 in. (300 mm) thick, should be placed and maintained at a minimum temperature of 55°F (13°C). Since this can lead to changes in mixture proportions and costs, it becomes important to discuss this at the prebid meeting. Once the project starts, concrete temperatures must be monitored at the plant on an hourly basis. Whenever there is a change in the measured concrete temperature by more than 5°F (3°C), the appropriate admixture type and dosage should be used so that the target setting times are achieved.

Effect of Temperature on Early-Age Strength

For certain structural elements, contractors would like to attain target in-place strength at early ages (2 to 7 days) to facilitate concrete operations like formwork release. In-place strengths are measured by nondestructive methods, such as maturity and pullout tests. Field-cured cylinder strength results are not accurate predictors of in-place strengths (Obla et al. 2008). When setting times are controlled by admixtures, early-age strengths are also controlled to some extent. However, early-age strengths are more influenced by the in-place concrete temperatures. If consistent in-place temperatures are maintained through suitable curing operations (insulation to trap the heat of hydration, external heat source, etc.), then consistent in-place, early-age strengths can be expected. However, this can become expensive under cold weather conditions with stringent early-age strength requirements. So, in such cases mixture proportions may have to be modified through the use of a lower w/cm, ASTM C150 Type III cement, and a lower dosage of supplementary cementitious materials. Since this can lead to changes in mixture proportions and costs, it becomes important to discuss this at the prebid meeting.

Effect of Temperature on Mixing-Water Demand

Past NRMCA research (Gaynor et al. 1985) showed that for every 10°F increase in concrete temperature, the average mixing-water demand for a consistent slump level at discharge of 4 ± 1 in. (100 ± 25 mm) and a delivery time of 60 minutes increased by about 5 lbs/yd^3 (1.8 kgm^3), which is about 2% of the mixing-water amount used typically. If this change in mixing-water demand is not taken into account it can lead to strength variations. Let us say for a particular mixture the measured mixing-water demand at 70°F (21°C) was 290 lbs/yd^3 (172 kg/m^3), including a Type A admixture dosage of 2 oz/cwt. (130 mL/100 kg). Column 4 of Table 9.2 shows how the mixing-water demand can vary with concrete temperature based on past NRMCA research (Gaynor et al. 1985). In order to achieve consistent mixing-water content and therefore consistent w/cm and strengths, it is suggested that the dosage of water-reducing admixtures be varied for every 10°F (5.5°C) change in the concrete temperature from the design temperature, which in this case is 70°F (21°C). At cool temperatures, if no water-reducing admixtures are used, cementitious content can be varied so that the same w/cm is maintained. This process is shown in Column 5 in Table 9.2. Doing this will also avoid the common observation of lower strengths during the summer and higher strengths during the winter for the same concrete mixture. Figure 9.2 shows the typical CUSUM chart for a mixture from a concrete plant (Bain and Obla 2007), which shows that as concrete temperatures increased, strengths decreased. Interestingly, the CUSUM chart suggests that slumps also went up, suggesting that there was a preference for higher slumps at higher temperatures, which can further increase the mixing-water content. Even though this discussion has been about concrete temperatures, higher ambient temperatures can further increase concrete temperatures, leading to even higher water demand.

VARIATION IN CONCRETE PERFORMANCE DUE TO DELIVERY TIME

Past NRMCA research (Gaynor et al. 1985) showed that when the delivery time was increased from 20 min to 90 min, the average mixing-water demand increased by 14 lbs/yd^3 (8.3 kg/m^3) and 21 lbs/yd^3 (12.5 kg/m^3) when the concrete temperature was maintained at 65°F (18°C) and 95°F (35°C), respectively. The increased mixing-water content was necessary to maintain a consistent slump level at discharge of 4 ± 1 in. (100 ± 25 mm). A 21 lbs/yd^3 (12.5 kg/m^3) difference in mixing-water content

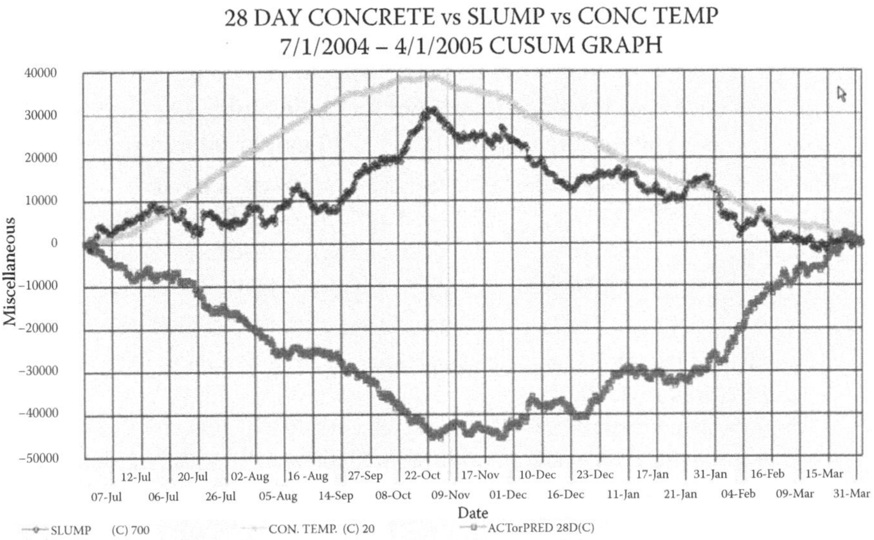

FIGURE 9.2 CUSUM graph showing strong dependence of 28-day compressive strength, and slump on concrete temperature over 12 months of concrete production at a plant. (Reprinted with permission from Bain, D., and Obla, K. H., "Concrete Quality Control—The Untapped Profit Center," *Concrete InFocus*, Fall 2007, Vol. 6, No. 3, NRMCA, pp. 63–69.)

can lead to a variation in compressive strength of over 400 psi (2.8 MPa) for a typical 0.50 w/cm mixture. This variation in delivery times occurs even for short travel distances due to traffic in urban environments and/or contractor schedules. Concrete producers should design their mixtures such that specified slump levels at the job site are achieved regardless of whether the delivery time is 20 min or 90 min. Producers can address this through one or a combination of the following ways:

1. Design mixtures so that there is no slump loss at all. This is impractical, but with suitable choice of retarding admixtures and higher amounts of certain SCMs, slump loss can be reduced; therefore, the variation in mixing-water demand can be reduced. Care should be taken that target setting times are still achieved.
2. Target slumps at the plant judiciously. Based on the past NRMCA research it can be estimated that the concrete slumps should be 1.5 in. (40 mm) and 2.5 in. (65 mm) lesser at 90 min than at 20 min for concrete temperature maintained at 65°F (18°C) and 95°F (35°C), respectively. Table 9.3 suggests target plant slumps. The plant target slumps vary over a narrower range than the specified limits at the job site and are selected on the following basis: The lower plant limit is selected so that even after a delivery time of 90 min the slump meets the lower limit of the specified slump range. The upper plant limit is selected such that even after a delivery time of 20 min the slump meets the upper limit of the specified slump range. Basically, this approach ensures that the plant target slumps are close to the upper limits

TABLE 9.3
Target Plant Slumps to Ensure Specified Slump Levels at the Job Site Are Achieved, Regardless of Whether the Delivery Time Is 20 min or 90 min

Concrete Temperature, °F (°C)	Plant Slumps for Specified Slumps of 5 ± 1.5 in. (125 ± 40 mm)
<80 (27)	5.0 to 6.5 in. (125 to 165 mm)
>80 (27)	6.0 to 7.5 in. (150 to 190 mm)

specified at the job site and are even higher when the concrete and ambient temperatures are high.

3. Use job-site admixture (water reducer) addition to compensate for slump loss provided qualified personnel are available to administer that. Automated admixture (water reducer) addition devices can also be considered when available.

SUMMARY

Establish a proper batching sequence to minimize the occurrence of head packs and cement balls and to improve uniformity of concrete.

For every truck, do an annual check for blade wear and internal buildup of hardened concrete.

Consider a mixing uniformity evaluation on truck mixers to establish critical levels of blade wear and buildup. The study should include at a minimum slump, air content, and compressive strength. Alternatively, perform a visual evaluation of slump during discharge. Trucks that show a problem with achieving uniformity of concrete should be removed from service and not used until the problem has been rectified. If the problem is not truck specific, it is likely to be due to a poor batching sequence.

Mixing speed should be selected so that the desired flow pattern is created inside the truck. Flow patterns can be studied at the same time random concrete uniformity studies are being conducted.

Monitor concrete temperature at the plant on an hourly basis. When setting time is a performance requirement for varying concrete temperatures, producers should establish admixture type (accelerator and retarder), dosage level, and mixture modification that would help to bring the concrete mixture to the target setting time. Whenever there is a change in the measured concrete temperature by more than 5°F, the appropriate admixture type, dosage, and concrete mixture should be used so that the target setting times are achieved.

Vary the dosage of water, reducing admixtures for every 10°F change in the concrete temperature from the design temperature to attain the same w/cm at different concrete temperatures.

Design concrete mixtures to minimize slump loss. Design concrete mixtures to attain slumps at the plant that are near the upper limit of the specified slump at the job site. When slump loss is encountered, use job-site admixture (water reducer) addition with qualified personnel.

10 Variation in Concrete Performance Due to Testing

This chapter discusses variability due to acceptance testing and methods to improve the quality of acceptance testing, thereby reducing the overall variability of concrete strength.

A MEASURE OF TESTING VARIABILITY

ACI 214R provides a means to judge the testing variability and is summarized below:

1. Variability due to testing is estimated by the within-batch coefficient of variation (V_1) calculated based on the average range (R) in strengths of companion (replicate) cylinders, comprising a strength test result, which are cast from the same composite sample of concrete tested at the same age using Equation 10.1. Since the cylinders are made from the same concrete sample, the material and manufacturing variability are assumed to be negligible, and the strength difference between the cylinders is assumed to be due to testing variability. Average range should be calculated from at least 10 strength test results. In Equation 10.2, the within-batch coefficient of variation (V_1), in percent, is determined from S_1 and the average strength (X).

$$S_1 = \frac{R}{d_2} \tag{10.1}$$

   where d_2 = 1.128, 1.693, 2.059 if the number of cylinders averaged for a strength test result are 2, 3, 4, respectively.

$$V_1 = \frac{S_1}{X} \times 100 \tag{10.2}$$

   The calculation of S_1 and V_1 for an example containing 10 compressive strength test results, is shown in Table 10.1.

2. Table 10.2, which is adapted from ACI 214R Tables 4.3 and 4.4, shows that for field construction testing the quality of testing can vary from Excellent (V_1 < 3%) to Poor (V_1 > 6%). The last row shows the calculated average

TABLE 10.1
Calculation of Within-Batch Variation (V_1,%)

Cylinder 1, psi (MPa)	Cylinder 2, psi (MPa)	Strength Test Result, psi (MPa)	Range, psi (MPa)	Range, %
6740 (46.5)	7120 (49.1)	6930 (47.8)	380 (2.62)	5.5%
7050 (48.6)	6750 (46.6)	6900 (47.6)	300 (2.07)	4.3%
5640 (38.9)	5830 (40.2)	5735 (39.6)	190 (1.31)	3.3%
5570 (38.4)	5550 (38.3)	5560 (38.3)	20 (0.14)	0.4%
6030 (41.6)	5700 (39.3)	5865 (40.4)	330 (2.28)	5.6%
5690 (39.2)	5650 (39.0)	5670 (39.1)	40 (0.28)	0.7%
5530 (38.1)	5600 (38.6)	5565 (38.4)	70 (0.48)	1.3%
5350 (36.9)	5320 (36.7)	5335 (36.8)	30 (0.21)	0.6%
4650 (32.1)	5080 (35.0)	4865 (33.6)	430 (2.97)	8.8%
5800 (40.0)	6080 (41.9)	5940 (41.0)	280 (1.93)	4.7%
Average		5837 (40.3)	207 (1.43)	

$$S_1 = \frac{207}{1.128} = 184 \text{ psi } (1.27 \text{ MPa})$$

$$V_1 = \frac{184}{5837} \times 100 = 3.2\%$$

TABLE 10.2
Standards of Concrete Control

Quality Standards (ACI 214)	Excellent	Very Good	Good	Fair	Poor
V,%	<3.0	3.0 to 4.0	4.0 to 5.0	5.0 to 6.0	>6.0
Calculated average Range of 2 Companion Cylinders that comprise a test result	<162	162 to 217	217 to 271	271 to 325	>325

Source: Adapted from ACI Committee 214R, "Evaluation of Strength Test Results of Concrete (ACI 214R-11)," American Concrete Institute, Farmington Hills, MI, 2011, 16 pp., Tables 4.3, 4.4.

range (R) of 2 companion cylinders, assuming a case where the measured average strength is 4800 psi (33 MPa). It is seen that the average range can vary from below 162 psi (1.1 MPa) (for excellent testing quality) to over 325 psi (2.2 MPa) (for poor testing quality). According to Ken Day (2011), R below 145 psi (1 MPa) is good, and exceeding 290 psi (2.0 MPa) suggests examination of the testing process. Comparison between different strength grades becomes possible when it is based on V_1 as opposed to R. To track V_1, concrete test reports need to report the strength of individual cylinders.

When test reports become available, the concrete producer could enter the results in a spreadsheet to calculate V_1 based on the last 10 strength test results. Luke Snell (1995) states that, if V_1 is between 4% and 6%, there are potential problems, and if V_1 is above 6%, the testing should be questioned. Realize that V_1 should be in the range of 2 to 3% due to expected variation in the ASTM C39 compressive strength test method. If it is a very low value, there may be cause to question whether cylinders are being tested to complete failure or if both cylinders are actually being tested.

3. Another way to measure testing variability is to use the precision statement in ASTM C39 that indicates the variation of companion cylinders—both 6 × 12 (150 × 300 mm) and 4 × 8 in. (100 × 200 mm) specimens in field and laboratory conditions. According to C39, 4 × 8 in. (150 × 300 mm) specimens tend to be a bit more variable. For example, C39 indicates that the range of strengths of two 6 × 12 in. (150 × 300 mm) cylinders made from the same sample by one technician should not exceed 8% of the average strength more than 1 time in 20. Tracking the range of cylinder results in a spreadsheet and evaluating whether this acceptable range is exceeded very often, indicates a problem with the testing. The calculation of within-batch range as a percent of the average strength for an example is shown in Table 10.1 (column 5).
4. To attain an overall concrete variation $S < 400$ psi (2.8 MPa) (corresponding to Excellent quality according to ACI 214R), testing quality should be at least Very Good ($V_1 < 4\%$).
5. V_1 does not measure the quality of testing in its entirety. Serious deficiencies in sampling and testing (initial and final curing, capping material, testing-machine calibration, etc.) would affect companion cylinders from the same sample equally and may not increase V_1 but still can lead to lower measured strengths and impact the standard deviation of strength test results. Let us take the case of initial curing, which has been identified as one of the main reasons for low strength test results. ASTM C31-10 states that for standard curing, immediately after casting, the cylinders shall be stored for a period up to 48 h (called *initial curing*) in a temperature range from 60 to 80°F (16 to 27°C) (68 to 78°F (20 to 26°C) for specified strength ≥6000 psi (41 MPa) and in an environment preventing moisture loss from the specimens. Also, the cylinders should be shielded from direct sunlight or radiant heating devices. Unfortunately, this is not always practiced in the field. It has been reported (Obla et al. 2005) that nonstandardized curing during the first 48 hours can cause a 1000 psi (7 MPa) loss in 28-day compressive strength for a typical 4000 psi (28 MPa) concrete mixture. Job-site conditions (temperature, humidity) and practices vary every day; therefore, cylinders cast at different time periods are expected to experience different initial curing conditions if not subject to standard curing. *However, companion cylinders cast from the same batch of concrete should experience identical initial curing conditions.* In summary, variations in initial curing conditions are unlikely to influence the range of compressive strength of companion cylinders prepared from a batch but will impact the batch-to-batch variation and

therefore the overall strength variability. Therefore, V_1 calculated according to ACI 214 does not take into account variations in initial curing. A similar argument can be made for delays and inappropriate means of transporting cylinders to the laboratory.

OTHER METHODS OF EVALUATING TESTING

As discussed previously, sometimes faulty testing will not contribute to higher V_1 but will still lead to low strength results and higher overall strength variability. Such instances can be identified if more than one of the following scenarios occur together.

Other Property Measurements

If low strengths are measured, the corresponding test results for concrete slump (ASTM C143), air content (ASTM C231 or C173), temperature (ASTM C1064), and density of fresh concrete (ASTM C138) or cylinder weights should be analyzed. High slumps may indicate excessive water addition, which can be the cause for the low strengths; high air contents will lead to strength reduction as well; high concrete temperatures will lead to higher water demands for the same slump. For given mixture proportions and materials, low concrete densities are primarily due to high air contents and/or water contents. Corresponding ASTM C917 reports of cement shipments can indicate whether the strength of the cement has decreased. If low concrete strengths are not accompanied by low concrete densities or low C917 cement strengths, then poor concrete strength testing can be a possibility.

Producer Testing

If the producer performs concrete strength testing at the plant, reduction in strength acceptance test results without a reduction in strength test results measured at the plant would suggest either poor job-site testing practices or increase in mixing-water content and/or air content at the job site. If concrete density measured at the job site or through cylinder weights at the time of demolding (or testing) does not show a decrease, then the mixing-water content and air content is unlikely to be changed, and therefore, the low strength is most likely due to poor job-site testing practices. Alternatively, internal tests performed by a technician with the concrete supplier at the job site can also be used to identify these problems.

7- to 28-Day Strength Gain

For a given mixture, the strength gain between 7 and 28 days is unique and should be relatively consistent. One should not assume that the typical rule of thumb that 7-day strength is 75% of 28-day strength applies for all mixtures. If the 7–28 day strength gain varies more than 40% of the measured average 7–28 day compressive strength gain of that mixture it may indicate a testing error. Cylinders that are exposed to cool temperatures during initial curing are likely to give lower 7-day test results and acceptable 28-day results as long as they are not frozen. Similarly, cylinders that are exposed to high temperatures during initial curing are likely to give higher 7-day test results and lower 28-day results. When test results are received, concrete producers should update a plot of the 7–28 day strength gain. Let us say on an average the 7–28 day strength gain for a mixture is 1000 (7 MPa) psi. If the strength gain is less than 600 psi (4.1 MPa) or greater than 1400 psi (9.7 MPa), the cause should be investigated.

Cylinder Density

Densities of the acceptance test cylinders should be measured to the nearest 0.1 lb/ft^3 (1.6 kg/m^3) at the testing laboratory by weighing the cylinders as soon as they are demolded and the surface moisture is wiped off. Cylinder weights could also be measured just before placing them in the testing machine. Ken Day (2006) states that if the average cylinder density range between companion cylinders exceeds 1 lb/ft^3 (16 kg/m^3), it indicates poor fabrication and testing techniques. Maximum range should not exceed 2 lb/ft^3 (32 kg/m^3). Measuring and monitoring cylinder density variation will also have the effect of raising the overall testing quality as well.

Laboratory Reports

It is necessary for the laboratory reports to report the maximum and minimum temperatures during initial curing and the duration of initial curing. Transportation times should also be recorded. In a past case study, cylinders made on Fridays had lower strength problems because the testing laboratory picked up the cylinders on Mondays, thus exposing the cylinders longer than 48 h in the field.

Ensure that all the reporting requirements of ASTM C31 and C39 are included in the test report. Review this information to possibly identify other causes. Ensure that the test results represent the correct date of pour and the correct project. Low test results observed on a specific date may indicate a different testing technician. Maintain a record of the test results in chronological order in the sequence they were made as the ACI acceptance criteria are based on consecutive test results. Keep a record of the ambient temperature and other weather conditions when tests were made. Compare these to the information on the test results.

ACI CODE AND SPECIFICATION REQUIREMENTS RELATED TO CONCRETE TESTING

Section 5.6 of the building code (ACI 318-11)[3] has the following requirements that can help improve testing quality:

1. Testing agency performing acceptance testing shall comply with ASTM C1077. Many jurisdictions require commercial laboratories to hold accreditation by an independent body.
2. Certified field and laboratory testing technicians shall perform all the testing.

3. Cylinders for strength tests shall be molded and standard-cured in accordance with ASTM C31. The reporting section of ASTM C31 states that for standard curing, the initial curing method with maximum and minimum temperatures should be reported.
4. All reports of acceptance tests shall be provided to the licensed design professional, contractor, concrete producer, and, when requested, to the owner and the building official. This is a new ACI 318 requirement. In many projects, concrete producers generally do not receive test reports and are informed only in the case of low-strength test results. This defeats one of the important purposes of testing standard-cured cylinders, which is ensuring the quality is adequate. If test results are available regularly in a timely manner (preferably as soon as the 7-day tests are conducted), the producer can notice trends, identify potential strength decreases earlier, and make mixture adjustments if needed. This will certainly reduce the time and expense incurred in low strength investigations. The 7-day strength tests do not have acceptance requirements, though.

Specification for Structural Concrete (ACI 301-10), which is adopted by the commonly used Construction Specification Institute's Masterspec, is consistent with the ACI 318 requirements and specifically states the following:

1. Field tests of concrete shall be made by an ACI Concrete Field Testing Technician Grade I or equivalent.
2. The contractor's responsibility is to provide space and source of electrical power on the project site for initial curing of concrete test specimens, for the sole use of the testing agency.
3. The testing agency will conduct concrete strength tests during construction by making and curing test specimens in accordance with ASTM C31 and testing them according to ASTM C39/C39M. The testing laboratory report should include detailed information about storage and curing of specimens before testing location in the work where the concrete represented by each test was deposited, date and time of sampling, and batch ticket number.
4. Test reports should be provided to all parties within 7 days after the tests are performed.
5. Density (ASTM C138) should be measured, along with slump, air content, and temperature, whenever cylinders are made for strength tests.

Specification for testing agency requirements (ACI 311.6-09) states that the testing laboratory shall make strength test specimens in accordance with ASTM C31. Specimens shall be stored under conditions that meet the requirements of ASTM C31 and shall be verified by a testing agency. Such storage shall have temperature controls to maintain ASTM C31 temperature requirements. Calibrated temperature recording devices shall be used to record daily maximum and minimum temperatures of the initial curing environment. The testing agency will transport concrete specimens in accordance with ASTM C31 and shall conduct final curing of strength test specimens in accordance with ASTM C31 and C511. The testing agency shall be accredited in accordance with the requirements of ASTM C1077 by a third-party agency.

STEPS TO IMPROVE THE QUALITY OF ACCEPTANCE TESTING

EDUCATION

On an annual basis, local ACI and engineering group lunch presentations/group discussions can be arranged to discuss ACI 318 Code, 301 and 311.6 specification requirements pertaining to the quality of testing. Articles that outline the impact of nonstandard curing and/or testing quality on compressive strength test results can be distributed (Meininger 1983; Goeb 1995; Obla et al. 2005; Richardson 1991). Testing laboratories can be invited to discuss their best practices for maintaining high testing quality. Table 10.3 provides a list of factors related to testing that can impact the strength of concrete.

For ready mixed concrete companies, truck mixer drivers are their primary "eyes" at the job site. Drivers should be trained on the test methods covered in the ACI Field

TABLE 10.3
Factors Related to Testing That Can Affect Strength Test Results

Factors	Details
Cylinder molds	Leaky molds, out-of-round molds, concave or convex top or bottom surface, lack of lid or plastic bag during initial curing
Sampling	Sample not representative of batch or not remixed before cylinder fabrication, incomplete consolidation leaving large voids, uneven/rough ends due to poor finish on top, lack of support on bottom
Initial curing	Temperature not between 60 and 80°F (16 and 27°C), drying, direct sunlight exposure, vibration damage too close to activity
Marking, transporting cylinders	Lack of proper record keeping concerning concrete batch (other property measurements, placement locations, time in field, etc.), cylinder marked improperly (not on lids), rough handling of cylinders, delays in moist curing at laboratory
Moist curing at laboratory	Temperature control between 70 and 77°F (21 and 25°C), maintenance of free water on all surfaces cylinders at all times, drying during capping operations
Capping	Proper temperature of sulfur melting pot, thick caps (>3/16 in. [4.75 mm]) due to uneven ends of specimens or rapid cooling of capping agent, cap bonding, use of reused and low strength capping material, caps at least 2 hours before breaking cylinders, caps aligned with cylinder axis, nonplane caps, and capping plate
Compression testing	Testing machine in poor mechanical condition or not calibrated, imprints of bearing blocks on the caps that show eccentric positioning, platens not stiff to too small, machine frame stiffness, top bearing block spherically seated so that it can rotate to come in full contact with top cap, moist specimen condition, rate of loading between 28 and 42 psi/s (0.2 and 0.3 MPa/s), cylinders broken to failure, measurement of specimen diameter to the nearest 0.01 in. (0.25 mm)

Source: Adapted from Richardson, D. N., "Review of Variables That Influence Measured Concrete Compressive Strength," *Journal of Materials in Civil Engineering*, Vol. 3, No. 2, May 1991, pp. 95–112.

Testing Technician Grade I certification and even obtain certification. They can then correct improper sampling, initial curing and fresh concrete testing practices by testing technicians, or document bad practices with cameras on cell phones (if permitted by the company). For example, according to ASTM C172, strength test cylinders should be made from a composite sample obtained from near the middle of the load. According to ASTM C94, preliminary samples (discharge of ¼ cu. yd. of concrete) are acceptable only for slump and air content adjustments.

Round-Robin Testing Programs

Several locations around the country conduct annual round-robin (RR) compressive strength testing programs. Basically, a RR program involves making a large set of cylinders from a concrete sample procured from a truck after discarding the first ¼ yd^3 0.2 m^3) of concrete that will typically not provide a representative sample. All the specimens are left in molds covered with plastic and stored in a 70°F (21°C) laboratory environment for the first 24 h. Following that, each participating testing laboratory picks up their cylinders, cures them in their laboratory according to ASTM C511, and tests them at 28 days. The results are sent to the coordinator of the RR program who gives a code number for each laboratory. The results are statistically analyzed and each laboratory is provided a rating. The rating criteria are summarized in Table 10.4 and are the same used by the Cement and Concrete Reference Laboratory (CCRL) in their proficiency sample testing program (Crandall and Blaine 1959). ASTM C1077 indicates that when a laboratory obtains low ratings (<2) in proficiency sample testing, they should document an investigation into the reason for the

TABLE 10.4
Laboratory Rating Calculated from Round-Robin Test Results

Laboratory Rating	Laboratory Z Score
5	≤1.0
4	>1.0 and ≤1.5
3	>1.5 and ≤2.0
2	>2.0 and ≤2.5
1	>2.5 and ≤3.0
0	>3.0

Source: Adapted from Crandall, J. R., and Blaine, R. L., Statistical Evaluation of Interlaboratory Cement Tests," *ASTM Proceedings*, Vol. 59, 1959, pp. 1129–1154.

$$\text{Laboratory Z score} = \frac{(\text{Laboratory test result} - \text{Overall average})}{\text{Standard deviation of results}}$$

Z scores can be negative in which case the laboratory ratings shall be negative. A negative rating indicates that the laboratory's result was below the average; a positive rating indicates that the laboratory's result was above the average.

low rating and indicate corrective action taken. An annual RR program is cosponsored by the Washington Area Council of Engineering Laboratories (WACEL) and the Ready Mixed Concrete Producers Technical Committee (RMC-PTC) serving the Washington, DC, metropolitan area. Figure 10.1 is a plot of the reported data (Lobo 2011). This type of program also establishes a multi-laboratory variation of

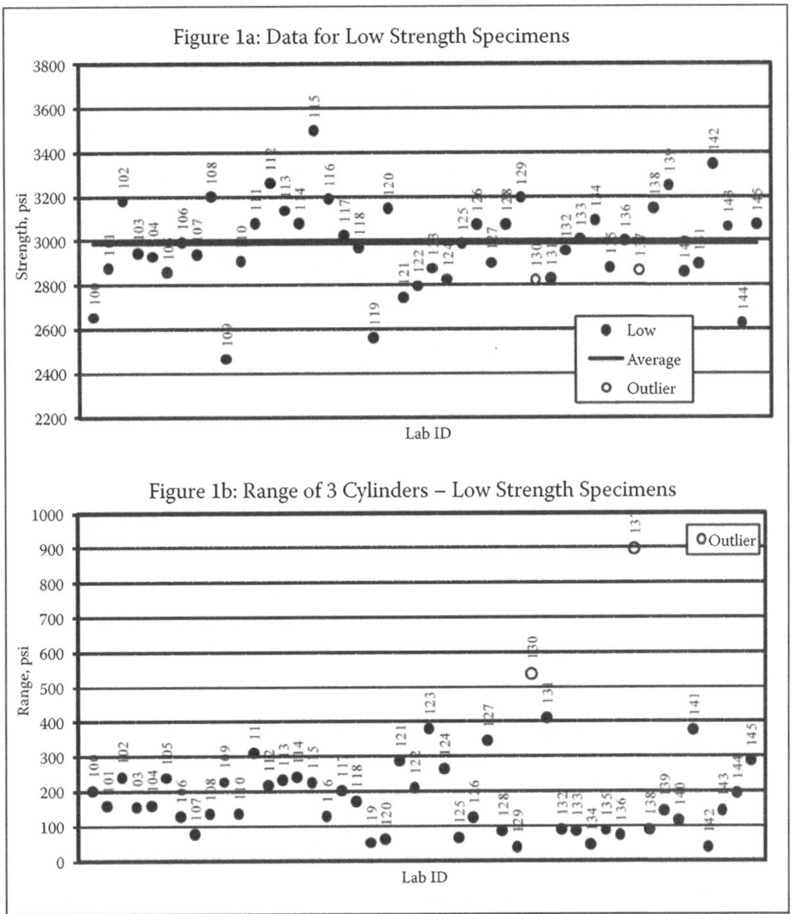

Number of cylinders	3		Average Strength	2986 psi
Number of labs	44		Maximum	3501 psi
			Minimum	2466 psi
Within test standard deviation		105.2 psi	(3 cylinders averaged for the test result)	
Coefficient of variation		3.5%		
Multi lab standard deviation		220.7 psi	Multi-lab "d2s"	618 psi
Coefficient of variation		7.4%		

FIGURE 10.1 Analysis of round-robin testing of laboratory strength test results. (Reprinted with permission from Lobo, C. L., "Report on the Round Robin Strength Testing Program-2011," Cosponsored by WACEL and RMC-PTC, June 2011, NRMCA, 9 pp.)

results as several laboratories test essentially the same concrete. This is useful to determine an acceptable difference between results of two different laboratories.

Incentives to Testing Technicians

Another approach that seems to work is to provide incentives to testing technicians who achieve good testing quality as measured by low within-batch coefficient of variation, V_1. Instead of V_1, average range and maximum range of companion cylinders can also be compared as long as the measured strengths are within ±1500 psi (±10 MPa). A concrete producer (Bain and Obla 2007) tracked the average and maximum range of companion cylinders of all technicians who tested concrete supplied by his company. At the end of the year, he identified the top two technicians who had the lowest average range and provided a gift certificate and a letter attesting to their achievement. These letters were proudly displayed on the notice boards of the testing laboratories at which the technicians worked. The producer stated that this practice motivated the laboratory and technicians to improve their testing standards. Another testing laboratory maintained a list of V_1 of all technicians and displayed them on the company notice board. The laboratory reported that this encouraged a healthy competition between technicians to try and do a better job and to attain low V_1. Laboratories are also known to tie pay incentives to the measured V_1.

Preconstruction Conferences

The NRMCA-ASCC checklist for the concrete preconstruction conference states that preconstruction meetings are very important as many potential problems can be avoided before the start of the project when the cost impact is low. The checklist allocates responsibilities and establishes procedures related to concrete construction, including acceptance testing. Some of the checklist items pertaining to initial curing of the acceptance test specimens and transportation to the laboratory have been included as Appendix D. The checklist also discusses procedures to be followed for evaluation of low strength testing such as nondestructive testing and core testing. The matter of who pays for the additional testing if the specified strengths are confirmed should be discussed. The standard practice is the producer does not pay if the specified strengths are confirmed by core tests (NRMCA Publication 133-11). Having this discussion at the preconstruction conference ensures that the initial curing is correctly done at the job site.

Other Strategies

Curing boxes are available that can maintain temperatures between 60 and 80°F (16 and 27°C) regardless of the ambient temperature. These can cost upward of $2000 and can accommodate about 20 cylinders at a time. Curing boxes require power, which is not always available at the job site. In smaller projects where it becomes cost prohibitive, in cold weather, insulated containers have been used (Detwiler et al. 2009), and if very low temperatures are expected, additional cylinder molds containing hot water have been placed inside the box. Similarly, during warm weather, water-filled ice chests have been used (Detwiler et al. 2009) with cylinder molds filled with ice utilized when ambient temperatures exceed 90°F (32°C). For warm weather, the most effective approach is to immerse the specimens in water with ice if necessary. Ozyildirim (2011) suggests the use of curing boxes with continuously recording thermometers. Goeb (1995) has suggested that for strength test results to be used in acceptance evaluation a mandatory certifying statement that the results were performed in accordance with ASTM C31 and C39 should be recorded on the test report.

SUMMARY

When concrete test reports become available, the concrete producer should enter the results in a spreadsheet to calculate V_1 based on the last 10 strength test results and the range of cylinder strengths from a single strength test result. If V_1 exceeds 4% or if the range exceeds the acceptable range (in the precision statement of ASTM C39), the testing procedures should be investigated.

A low within-test variation, V_1, below 4% or a range below the value stated in the ASTM C39 precision statement is not an assurance of good testing procedures. Many sampling and testing deficiencies will not contribute to higher V_1 but will still lead to low strength results and higher overall strength variability. Such sampling and testing deficiencies can be identified if combinations of the following scenarios exist:

1. Low concrete strengths not accompanied by low concrete densities, low C917 cement strengths, or producer's internal testing at the plant or job site.
2. Strength gain between 7 and 28 days varies by more than 40% from the average and no similar change in C917 cement strengths.
3. Average density range between companion cylinders exceeding 1 lb/ft^3 (16 kg/m^3) and maximum range exceeding 2 lb/ft^3 (32 kg/m^3).
4. Laboratory reports indicating typographical errors and nonstandard initial curing, such as specimens left for greater than 48 h in the field, temperatures outside the 60°F to 80°F (16°C to 27°C) range.

The following steps are suggested to improve testing quality:

1. Each locality should arrange for periodic meetings or seminars that include all stakeholders on how to improve testing quality with a focus on ACI 318 Code, 301 and 311.6 specification requirements such as requirement of certified laboratories, and technicians; distribution of test reports to all stakeholders within 7 days; responsibility of contractor for providing facilities for initial curing at the job site, and responsibility of testing laboratory for making, curing, and transporting test specimens in accordance with ASTM C31.
2. Ready mixed concrete drivers should be trained to spot, correct, or document nonstandardized sampling and testing procedures.
3. Each locality should conduct strength round-robin testing programs.
4. Producers or testing labs should measure testing variation and provide incentives to technicians who achieve good testing quality.

5. Preconstruction conferences should discuss responsibilities and procedures for initial curing and payment for additional testing in the case of low acceptance strength test results.
6. Curing boxes can help ensure initial curing is appropriate. For smaller projects, less expensive options for adequate initial curing has been suggested.

11 Internal Concrete Testing

Internal concrete testing could be done at the job site or at the concrete plant. If internal testing is done at the job site, it allows for a comparison of testing practices between the producer and the third-party testing laboratory. Internal testing at the ready mixed concrete plant as a means to measure and reduce concrete variability is discussed in this chapter.

A variety of ingredients such as cement, supplementary cementitious materials (SCM) like fly ash, slag cement, aggregates, water and chemical admixtures are used in concrete. Concrete producers use a wide range of methods to measure or quantify the variability of ingredient materials, including reliance on testing performed by material suppliers. Preceding chapters have addressed how concrete producers can develop relations with cement, fly ash, and aggregate suppliers and use timely material test data to make adjustments to concrete mixtures to take into account the material variations. It is important that concrete producers conduct regular concrete testing at the plant so that the cumulative material, manufacturing, and testing variation at the plant can be determined.

WHY TEST AT THE PLANT WHEN WE CAN GET JOB-SITE TEST DATA?

Job-site testing, typically performed by independent laboratories for acceptance, includes delivery-time variations that can lead to further variation in mixing-water content and air content. As discussed in preceding chapters, even though the influence of delivery-time variations on concrete performance can be minimized, it cannot be eliminated. Poor job-site testing practices, such as sampling, cylinder fabrication, improper initial curing, and transportation to the lab, can further add to the overall variations. In summary, variability of concrete indicated by job-site tests includes several factors that the concrete producer does not control.

Variations in the concrete tested at the plant help the producer pinpoint the causes of variation over which they have more control. It also helps the producer compare different plants, understand the best practices at plants that have low variation, and establish best quality practices at all the plants.

CRITERIA FOR PLANT TESTING

Selection of Mixture Classes

Concrete producers should identify one to two mixture classes at each plant that would cover the broadly different concrete types and materials at the plant. Typically, all concrete mixtures belonging to a similar strength level (±1500 psi [±10 MPa]) and air entrainment, using the same cement/SCM type and aggregate type (crushed or rounded), and coming from the same geological origin are considered as belonging to a mixture class. Mixture classes are clearly not the same as actual mixtures, and several mixtures can belong to a mixture class. The main aim of the producer is to select just one or two mixture classes that will be produced several times during the week at a concrete plant. In Australia and the United Kingdom, testing performed by concrete producers are used for determining the acceptability of concrete (AS 1379, 1997; EN 206-1, 2000). Since our purpose is quality assurance as opposed to acceptance testing, it is suggested that concrete producers use the following methodology to select one to two mixture classes for each plant.

Slump: The selected mixture class (es) should have target average slumps between 3 and 7 in. (75 and 175 mm).

Air entrainment: Select one nonair-entrained concrete mixture class. In addition, select one air-entrained concrete mixture class if any air-entrained concrete is produced at the plant.

Aggregate type and origin: If the plant produces concrete with various aggregate types (crushed versus natural) and geological origins, the selected mixture classes should utilize the most commonly used aggregate type and origin.

Cement type and source: If the plant produces concrete with various ASTM C150 portland cement types and sources, the selected mixture classes should utilize the most commonly used cement type and source.

SCM type and source: If the plant produces concrete with various SCM types, sources, and SCM blends (not dosage), the selected mixture classes should utilize the most commonly used SCM type, source, and blend.

Admixture type: The selected mixture class (es) should contain the most common admixture type/dosage used in the plant.

Strength level: The selected mixture class (es) should have a specified strength between 2500 and 5000 psi (17.5 and 35 MPa).

The above discussion is summarized in Table 11.1. A concrete producer in a northern climate may be faced with the following:

- Air-entrained concrete is produced at the plant.
- Out of the concrete produced, 55% has fly ash, 20% has portland cement only, 15% has slag cement, and 10% has fly ash + slag cement blends.

Based on the discussions above, the producer should choose two mixtures classes—fly ash mixture with and without air entrainment.

Sampling and Types of Testing

After thorough mixing in the truck at least 1 ft^3 (28 liters) of concrete should be discharged in a wheelbarrow. The sample should be ideally obtained after discharging the first ¼ yd^3 (0.2 m^3) of concrete that will typically not provide a representative

TABLE 11.1
Selection of Mixture Classes

Variable	Mixture Classes
Air entrainment	One nonair-entrained concrete mixture One air-entrained concrete mixture if any air-entrained concrete is produced at the plant
Slump	Target average slumps 3 to 7 in. (75 to 175 mm)
SCM type and source	If the plant produces concrete with various SCMs, select mixture class with the most commonly used SCM type, source, and blend
Cement type and source	Most commonly used cement type and source
Aggregate type and origin	Most commonly used aggregate type and origin
Admixture type	Most common admixture type/dosage used in the plant
Strength level	Specified strength between 2500 and 5000 psi (17.5 and 35 MPa)

TABLE 11.2
Types of Testing

Types of Mixture Classes	Types of Testing
All	Slump (C143), density (C138), and temperature (C1074)
Air-entrained only	Air content (C231 or C173)
Nonair-entrained only	Compressive strength (C39)

Note: After thorough mixing in the truck, the sample should be taken prior to any water addition at the slump rack.

sample. The sample should be taken prior to any water addition at the slump rack. The following tests are suggested and summarized in Table 11.2:

For all mixture classes: slump (ASTM C143), density (ASTM C138), and temperature (ASTM C1064).
For air-entrained concrete mixture classes: air content (ASTM C231 or C173 if using lightweight aggregates or aggregates with high absorption).
For nonair entrained concrete mixture classes: compressive strength (ASTM C39).

Compressive strength tests should be conducted at 28 days and if possible at 7 days. The average of two 4 × 8 in. (100 × 200 mm) cylinders is taken as the compressive strength test result at that age. The primary aim of strength testing is to track ingredient material variability and its effect on concrete quality as measured by the strength variation. In air-entrained concrete, the variation in air content between batches alone can significantly influence the compressive strength variation. Therefore, compressive strength is not tested for air-entrained concrete mixtures.

On each occasion, the suggested sampling and testing should not take more than 10 min, and there is no need to retain the mixer truck during that period.

TABLE 11.3
Frequency of Testing

Average Weekly Production at Plant, yd³ (m³)	Frequency of Testing for a Week of Production: Slump, Density, Temperature, and Air Content (if applicable)	Compressive Strength
<400 (306)	2[a]	1
400 to 800 (306 to 612)	5[a]	2[a]
>800 (612)	10[b]	4[a]

Note: Compressive strength testing is done only for the nonair-entrained concrete mixture class, and air-content testing is done only for the air-entrained concrete mixture class. At the time of testing, any change in material shipments since the previous test should be noted.

[a] On different days of production.

[b] No more than 2/day of production.

Frequency of Testing

The next step is to decide how often to test the selected mixture classes and the type of tests to be conducted. This will depend, to some extent, on the personnel resources available at different plants. Ideally, it is desirable to do concrete testing every time a new material shipment is received. However, that may happen several times/day in some high volume plants and only a few times/month in low volume plants. So the frequency of testing is suggested in Table 11.3, and it depends on the weekly concrete production at the plant. Since compressive strength testing is done only for the nonair-entrained concrete mixture class, the stated frequency for compressive strength testing in Table 11.3 is the same as that for the nonair-entrained concrete mixture class. Slump, density, temperature, and air content are tested more often since they include testing of the air-entrained concrete mixture class as well. Testing should be conducted as early in the day as possible after avoiding the first two batches of the day. At the time of testing, any change in material shipments since the previous test should be noted.

DATA ANALYSIS

Control Charts

Plotting the results on control charts is an effective and visual means of evaluating data. Separate control charts should be plotted for each property—slump, air content, density, air-free density, temperature, and compressive strength. Control limits can be established to trigger an investigation if any result falls outside these limits. For each property, test results belonging to all mixture classes should be plotted on a single control chart. Since the mixture classes may have different targets of that property, it is suggested that the control chart be plotted as measured property minus the target average value. This target value may be from a trial batch or calculated from at least 30 test results of that mixture class from a past project. Once at least 15 test results are available, the target average value for that mixture class can be calculated and used for subsequent data. In this manner, all mixture classes can be plotted on a single control chart for that property, which helps to make the analysis in a shorter duration. Control charts should also be plotted separately for each mixture class. This helps to identify if only a certain mixture class exceeded the control limits frequently. It can also help assign causes for these changes. By drawing the control charts over an extended period (up to 1 year), investigating results that exceed control limits and making changes to quality practices, it is possible to maintain slump, air content, density, and compressive strength within the control limits, thereby ensuring low-variability concrete. In the beginning, the control limits may be frequently exceeded. However, with time, this should reduce. After 6 months of testing, if no improvements are observed, it could be due to the following reasons: poor choice of mixture classes, insufficient testing, inadequate analysis, causes not properly identified, or failure in adopting the suggested improvement in quality practices.

Slump

Control limits of ±1.5 in. (±40 mm) are suggested for slump. If the slump measured exceeds the control limits, it suggests that either the mixing-water content is not within tolerance or the mixing-water demand for that mixture has changed due to a change in characteristics of the materials, concrete temperature, or air content. A step-by-step investigation as outlined earlier in the chapter on mixing-water control should be carried out to understand the underlying cause. Frequency of testing should be increased to every load that belongs to either one of the mixture classes until the mixing-water content is determined to be in the target range, and any change in mixing-water demand is addressed by a change in water-reducing admixture dosage or by a change in other factors, such as characteristics of the materials, batched quantities, concrete temperature, or air content.

A control chart with 30 consecutive test results is shown in Figure 11.1. For the first 15 data points, the slump results exceed the control limits. It varies within ±3 in. (± 75 mm), and between points 16 and 30, the slump results vary within ±1 in. (±25 mm) suggesting that better mixing-water control is being practiced from point 16 onward.

Air Content

Air content tests are conducted only for the air-entrained concrete mixture class. Control limits of ±1.5% are suggested for air content. If the air content measured exceeds the control limits, it suggests that the AEA dosage has changed or AEA demand has changed due to a change in characteristics of the materials, batched quantities, concrete temperature, or slump. Causes for this should be investigated systematically as follows:

1. Was the air-entraining admixture dosage out of tolerance by more than ±20%? According to ASTM C94, admixture dosages are supposed to be within ±3% of the batch admixture amount or ± the dosage required for 100 lb (100 kg) of cementitious materials, whichever is greater. Batching

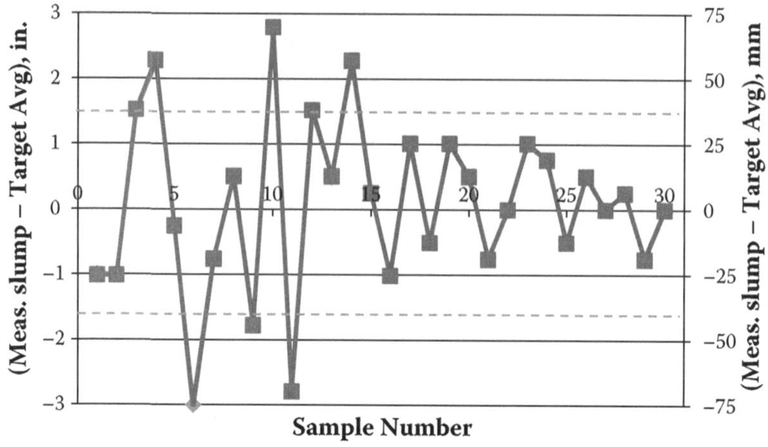

FIGURE 11.1 Control chart of slump test results.

accuracy of dispensers of liquid admixtures should be checked at least every 6 months. Periodically, it may be desirable to collect admixture at the point of discharge into the mixer to verify the accuracy of the dispensing system. This process checks the integrity of the lines through which the admixture flows in addition to the dispenser meter accuracy.

2. Is the air content test apparatus out of calibration?
3. Did the corresponding slump test result also exceed control limits? If two truckloads are batched one after the other and the only difference between the two is a difference in mixing-water content (greater than 10 lb/yd^3 [6 kg/m^3]), the truck load with lower mixing-water content is likely to have lower slump and lower air content.
4. Was the concrete temperature more than 10°F different from the concrete temperature than the mixture was designed for? Higher concrete temperatures will require greater amounts of air-entraining agent to attain a given air content.
5. Was the batched quantity of any of the solid materials varying by more than 5% of the target values?

If the answer to these questions is no, then changes in material characteristics, such as excessive fines in aggregate, dust on the aggregate surface, variations in fly ash, and other cementitious shipments should be evaluated. The AEA dosage should be adjusted suitably, and frequency of testing should be increased to every load until the air content falls within the control limits.

Density

Control limits of ±2 lb/ft^3 (± 32 kg/m^3) are suggested for density. If the density measured falls outside the control limits, possible reasons can be due to quantity of mixing-water content, air content, changes in aggregate relative density (specific gravity), or significant differences in actual batched quantities. Mixing-water content and air content will vary from batch to batch. A low density along with a high slump may be indicative of higher mixing-water content. If slump is the same or low, a high air content by itself could have contributed to the low density. Since the density test helps identify changes in mixing-water content and air content, it should be done every time the slump and air content tests are done.

Air-Free Density

Another parameter that can be used to check whether the mixture composition is similar to the mixture as designed is the air-free density. This is calculated as follows:

$$\text{Air-free density} = \frac{\text{Measured density}}{(1 - \text{Measured air content})}$$

Control limits of ±1.5 lb/ft^3 (±24 kg/m^3) are suggested for air-free density. If the air-free density falls outside the control limits and recorded batch quantities are reasonably in line, it suggests that most likely the mixing-water content was not correct. This could be due to excess water in the mixer before it was charged, incorrect aggregate moisture content, or inadvertent addition of water after the mixture was batched. In typical concrete mixtures, a 1.5 lb/ft^3 (±24 kg/m^3) difference in air-free density is caused by a 30 lb/yd^3 (18 kg/m^3) change in mixing-water content, which is enough to cause a 600 to 750 psi (4 to 5 MPa) strength difference. However, the density results alone falling outside the control limits should not trigger an investigation.

Temperature

As discussed in an earlier chapter on variation due to manufacturing, concrete temperature has a significant influence over slump/mixing water demand, air content, and set time. Therefore, concrete temperature control charts should be closely analyzed with the other control charts. Two sets of control limits are suggested at ±5°F (±3°C) and at ±10°F (±6°C). If the ±5°F (±3°C) limit is exceeded, setting time can be affected, thereby requiring change in mixture proportions if consistent set times are expected for a concrete application that needs finishing, such as slabs. If the ±10°F (±6°C) limit is exceeded, mixing-water demand and air content can be affected, and therefore mixture adjustments may be needed to ensure concrete of consistent workability, mixing-water content, air content, and strength.

Compressive Strength

Compressive strength tests are conducted only for the nonair-entrained concrete mixture class. Control limits for strength are recommended at ±2 standard deviations (S). The S used should be the target S, which may be known from past project. If that is not known, a standard value of, say, 500 psi (3.5 MPa) can be chosen. While strength acceptance criteria are set on the low end, a two-sided control limit is useful to evaluate reasons for excessively high strength, since this can be an economic issue for a project where excessive strength can impact the profit margin for the company. Assuming that concrete compressive strength follows a normal probability distribution, the following are the associated probabilities of occurrences of strength test results that can be used to trigger evaluation for assignable cause and initiate corrective action:

a. 4.5% probability that a 28-day strength test result will fall outside the control limits
b. 0.8% probability that 7 consecutive test results are on one side of the average
c. 0.2% probability that two consecutive test results fall outside the control limits
d. 0.05% probability that two consecutive test results fall outside the control limits on the same side

Primary causes for strength variations are variations in the characteristics of cementitious materials, organic impurities in fine aggregate, coarse aggregate dust/bond, mixing-water content, air content, cementitious content, mixing, sampling, curing, and testing.

If the 28-day strength measured falls outside the control limits, a systematic investigation should be conducted as follows to understand the underlying cause with a view to reduce that in future:

1. Did the corresponding density test result also fall outside the control limits? A low compressive strength in conjunction with a low density will suggest a high mixing-water content and/or high air content. Air content of nonair-entrained concrete mixtures should not vary by more than 1%, which means that the resulting strength variation is likely to be below 5%. A high slump and/or high temperature can also suggest high mixing-water content.
2. Were the batch weights out of tolerance—particularly, was the cementitious contents low or water content high?
3. Does the 7-day C917 data from the cement manufacturer show a similar strength change suggesting a change in the quality of the cement? The cement manufacturer conducts C917 tests 10 times a month and typically reports the most recent five-test moving average. It is important to ensure that the reported average includes the time corresponding to the production date of the cement shipment received at the concrete plant. This may be hard to ensure if cement is procured from a terminal.
4. Was there a change in fly ash (if used) fineness or 7-day strength activity index (SAI) data reported by the fly ash manufacturer? The reported results should be on samples corresponding to the production date of the fly ash shipment received at the concrete plant.
5. Has there been any change in concrete sampling and testing techniques and personnel? Are the within-batch ranges acceptable?

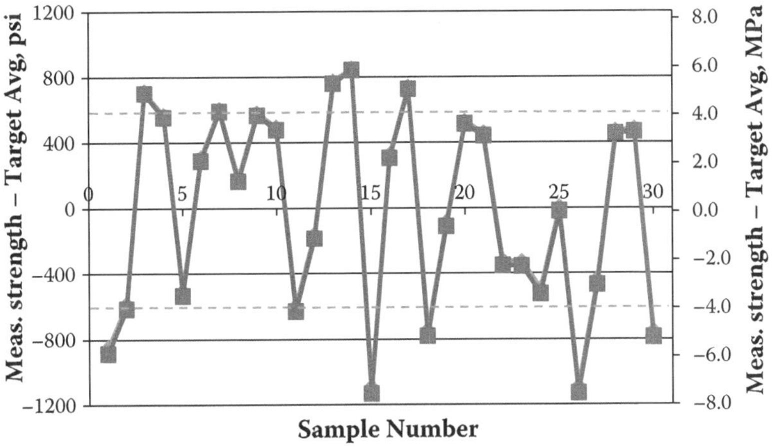

FIGURE 11.2 Control chart of compressive strength test results.

A control chart for 30 consecutive compressive strength test results is shown in Figure 11.2. The target standard deviation at the plant is set at 300 psi (2 MPa), which corresponds to excellent quality control for nonair-entrained concrete. The dotted lines correspond to two standard deviations above and below the target average strength. If the system is in control, only 5% of the test results should fall outside the two-control limits. A quick look shows that 10 test results (33%) fell outside the limits clearly suggesting that the actual standard deviation at the plant was much higher than the target value of 300 psi (2 MPa). In fact, it can be calculated to be 600 psi (4.1 MPa)!

In addition to the above analysis of the 28-day strength test results, testing and analysis can also be done on 7-day compressive strength test results. These results can help predict 28-day strengths and make small adjustments to mixture proportions so that the 28-day strength test results stay within control limits. However, it is suggested that the producer spend at least 12 months doing the analysis discussed here and start producing consistent concrete before resorting to mixture adjustments based on 7-day strength test results.

CUSUM Charts

Figures 11.3 and 11.4 are corresponding CUSUM charts for the slump, and strength test data plotted in Figures 11.1 and 11.2, respectively. Even though the control charts do not show trends, the CUSUM chart for strength clearly shows a rising trend from point 1 to 10, and then a decreasing trend from point 21 to 26. Similarly, the CUSUM chart for slump shows a rising trend from point 11 onward. CUSUM charts for 7-day strengths can be used to quickly spot trends, predict 28-day strengths, and make adjustments to mixture proportions as necessary. A study of CUSUM charts of one property should not be done in isolation. It should be combined with

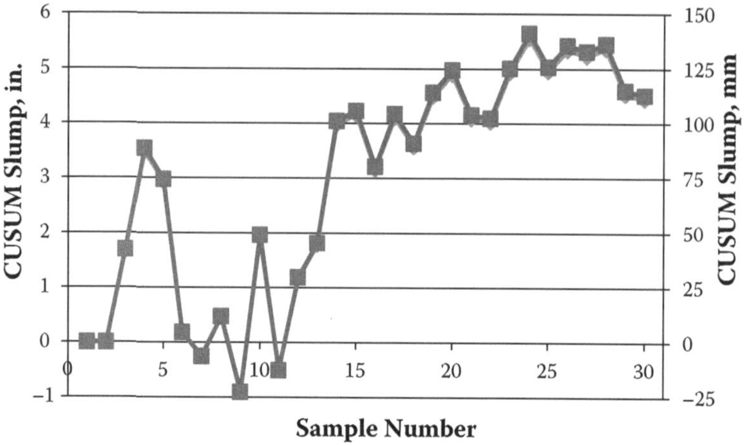

FIGURE 11.3 CUSUM chart of slump test results.

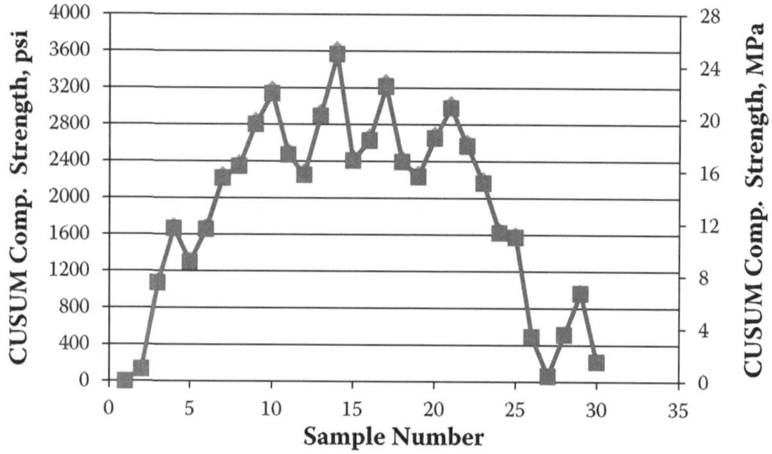

FIGURE 11.4 CUSUM chart of compressive strength test results.

CUSUM charts of other properties. For example, if CUSUM charts show rising slump, rising temperature, rising air content, and decreasing density, it would point to lower strengths. This approach is called multi-variable CUSUM and is similar to the concept described earlier of looking at control charts of various properties at the same time to understand the underlying cause of the variation. Further concrete strengths of different grades can be included in the same CUSUM chart for strength, which facilitates rapid decision making; the overall approach has been called by Day (2006) as Multigrade Multivariable Quality Control (MMQC). Day says that mixture adjustments with a view to reduce the strength should generally await confirmation from 28-day strength test results, but mixture adjustments to increase the strength should be made on low 7-day strength test results or even low-density test results, particularly if confirmed with high slumps and/or air contents.

SUMMARY

Conduct regular concrete testing at the plant. Use the data to compare plants and to better understand the causes of concrete variation and ways to reduce it.

Select 1 to 2 mixture classes at each plant that would cover the broadly different concrete types and materials at the plant. One mixture class should be nonair-entrained concrete, and the other mixture class should be air-entrained concrete if it is produced at the concrete plant. The sample should be taken after thorough mixing prior to any water addition at the slump rack. Slump, temperature, and density should be measured for all mixture classes. Air content should be measured only for air-entrained concrete mixture classes. Compressive strength should be measured only for nonair-entrained concrete mixture classes. Frequency of testing varies between 2/day and 1/week, depending on the average weekly plant production and the type of testing. At the time of testing, any change in material shipments since the previous test should be noted. On each occasion, the suggested sampling and testing should not take more than 10 min., and there is no need to retain the mixer truck during that period.

For each property (slump, etc.), test results belonging to one or both mixture classes should be plotted on a single control chart. Control charts for each property should also be plotted separately for each mixture class. Control limits are suggested for each property. Test results that exceed control limits should be investigated methodically. As part of this investigation, control charts of various properties should be viewed together. The key is that the investigation should lead to improved quality practices that help to maintain slump, air content, density, and compressive strength within the control limits, thereby ensuring low-variability concrete.

Once the random variations are reduced, it may be useful to do mix adjustments based on density or 7-day compressive strength results so that the 28-day strength variation is reduced and low breaks are avoided. CUSUM charts are very effective in data analysis.

12 Using Job-Site Test Results for Improving Concrete Quality

This chapter discusses the use of acceptance test results from testing laboratories as a means to quantify the variability, identify its main causes, and reduce it.

ACCEPTANCE TEST RESULTS

The annual NRMCA Industry Data Survey results indicated that between 2006 and 2010 a ready mixed concrete plant in the United States produced an annual average of 50,000 yd^3 (38,000 m^3) of concrete. Assuming that, on average half of that concrete is supplied to projects that include job-site acceptance testing at a frequency of one in every 150 yd^3 (115 m^3; minimum requirement per ASTM C94), it can be calculated that a concrete plant would receive about 14 project data sets/month. The test data are likely to be from different mixtures and/or projects. These data typically include slump, air content, temperature, density, and compressive strength at two ages (typically, one cylinder at 7 days, and two cylinders at 28 days). The density test results are useful for quality assurance and have been required in ASTM C94 and more recently incorporated in ACI 301-10.

DATA ANALYSIS

Data analysis should be conducted the same day the acceptance test data are received. Slump, air content, temperature, density, and strength test results that include 7 and 28 days, or other ages, should be input into the data analysis record as soon as the tests are available. In addition, other useful data to track include material sources or types, target quantities of mixture ingredients, recorded batch quantities, especially cementitious materials and water, and any in-house testing performed at the plant or job site, C917 cement strength data from the cement supplier, fly ash fineness and strength activity index, aggregate grading or fineness modulus of sand, and within-batch range of cylinder weights (if available). Data analysis can be done on spreadsheets, but it is useful to develop a visual means of evaluating trends in strength data by using control charts and/or CUSUM charts. Procedures for data analysis have already been discussed.

Rejecting Outliers

Can an acceptance test result be considered as an outlier? Let us say for a concrete project with a specified strength of 6,000 psi (41 MPa), test results were consistently above 6000 psi (41 MPa) and were averaging 6800 psi (47 MPa). The standard deviation of the test results was calculated as 600 psi (4.1 MPa). Suddenly, there is a single test result of 3700 psi (25.5 MPa). If all the test results formed a normal probability distribution, it can be statistically shown that there is only a 1 in 8.5 million chance that the strength test result will be 3700 psi (25.5 MPa) or lower! Is that grounds for rejecting that test result? The answer is no. Because if the producer had accidentally dispatched a different concrete mixture with a specified strength of 3500 psi (24 MPa), clearly the test result is valid and low-strength investigations should occur. In fact, strength test results should not be rejected unless there is clear evidence that the fabrication, curing, or testing was not in accordance with ASTM C31 and C39 requirements.

On the other hand, at least two cylinders made from the same sample of concrete are averaged to form a single test result. If one out of the three cylinders tested from the same sample broke at a much lower strength than the other two cylinders, can that be rejected? Once again, if there is clear evidence that the fabrication, curing, or testing was not in accordance with ASTM C31 and C39 requirements, that cylinder can be rejected without further analysis. Density is also a good measure. For some reason, if the density of that cylinder is substantially lower, then it may suggest poor consolidation. However, if such evidence is lacking, a statistical approach can be adopted. Since all three cylinders are made from the same sample, their strengths should not vary significantly. How much of a variation is acceptable? Referring to ASTM E178-08, ACI 214R states that generally the result from a single specimen in a set of three or more specimens can be discarded if its deviation from a test average is greater than three times the within-batch standard deviation (S_1). For $V_1 = 4\%$ and $\overline{X} = 4500$ psi (31 MPa), S_1 can be calculated as 180 psi (1.25 MPa). So, if a single specimen varied by more than 540 psi (3.72 MPa) from the test average, then that cylinder strength result can be rejected. If one specimen is rejected, the test average should be computed from the remaining specimens.

According to ACI 318, if a low strength test result is confirmed and calculations indicate that the load-carrying capacity is significantly reduced, three cores shall be taken for each low strength test result. Concrete in an area represented by core tests is considered structurally adequate if the average of three cores is equal to at least 85% of f'_c and if no single core is less than 75% of f'_c. Additional testing of cores extracted from locations represented by erratic core strength results are permitted. Since three cores are taken, if one of the core strength results is found to significantly vary from the other two core strength results, an outlier investigation as discussed in the previous paragraph can be conducted. The within-batch standard deviation can be calculated by assuming a higher coefficient of variation as core test results tend to be more variable. If the core strength result was found to be an outlier, a case can be made to extract another core from the same area.

Control Charts

Separate control charts should be plotted for each measured property: slump, air content, density, air-free density, temperature, compressive strength, 7-to-28-day compressive strength gain, and within-batch range of 28-day strength results. For each property test, results belonging to all mixtures should be plotted on a single control chart by subtracting the target average value from the measured property as discussed in Chapter 11, Internal Concrete Testing. The within-batch range does not have a target average value and is simply reported as a percent of the measured compressive strength test result.

Control Chart Limits

Table 12.1 lists the suggested control limits for the various properties. For typical mixtures, the control limits for density are chosen to be consistent with that of the air content.

The steps to be taken when control limits are exceeded have been discussed earlier in Chapter 11, Internal Concrete Testing. It is worth repeating here that by looking at the control charts of various variables at the same time, the underlying cause can be more easily determined. For example, if a high slump is accompanied by acceptable air content and low density, it may point to high mixing-water content. The steps to be taken when control limits are exceeded for 7-to-28-day compressive strength gain and within-batch range of 28-day strength results are discussed below.

TABLE 12.1
Limits for Control Charts for Various Properties

Property	Control Limits
Slump, in. (mm)	±1.5 (40)
Air content	±1.5%
Temperature, °F (°C)	±10 (5.5)
Density, lb/ft^3 (kg/m^3)	±2.0 (32)
Air-free density, lb/ft^3 (kg/m^3)	±1.5 (24)
Compressive strength	±2S
7–28 comp. strength gain	±40%[a]
Within-batch range	8%[b]

[a] Expressed as percent of the measured average 7–28 compressive strength gain.

[b] This is the acceptable range for two 6 x 12 in. (150 × 300 mm) concrete cylinders as reported in ASTM C39 precision statement. For 4 × 8 cylinders and for range calculated from three cylinders, refer to the ASTM C39 precision statement.

If the 7-to-28-day strength gain falls outside the control limits, it indicates one of the following: a testing error, variation in handling and curing of test specimens, variation in characteristics of cementitious materials, or incorrect batching of mixtures (i.e. batching a mixture with a significantly different dosage of SCM than designed). If tests done on other concrete samples around the same time period or the corresponding C917 reports of cement shipments do not show any changes in the rate of strength gain, and if there is no major batching error in SCM dosage, then it is likely that there is either a testing error or curing variation of test cylinders.

If the within-batch range falls outside the control limits more than 5% of the time (i.e., 1 in 20 results), then the testing quality needs to be investigated. A different way of evaluating the testing quality is to maintain a moving average within-batch coefficient of variation (V_1) calculated from the past 10 consecutive strength test results. If $V_1 > 4\%$, testing quality needs to be investigated. Chapters 7 and 10 show the detailed calculations in the development of these charts. As discussed in the previous chapter on testing, these evaluations of testing quality, however, may not capture the variability introduced due to variations in curing.

A key factor with evaluating data is to catch the negative trend as soon as possible so corrective action can be taken. For strength data, an evaluation for cause should be started as soon as more than one 7-day strength test result falls outside the control limits. This ensures that strength downturns and quality problems are discovered early on and addressed, thus reducing problems that may be more expensive at a later point. Typically, many 7-day results are available before 28-day results, and so a pattern can be identified early on. At the very least, an investigation should be done as soon as the 28-day strength test results fall outside the control limits to ensure that the causes can be identified and thus future variations in strength can be reduced.

Monitoring S of Compressive Strength

The 15 test moving S at the plant should be monitored when at least 15 test results are available. Since S is likely to be similar for several concrete mixtures with a specified compressive strength ≤5000 psi (35 MPa), test data from different mixtures can be included. S of air-entrained concrete might be kept separate from nonair-entrained concrete, and strength data for specified strength >5000 psi (35 MPa) can be evaluated separately also. S can also be calculated from the average range of successive pairs of test results for the same mixture. At least 15 sets of successive pairs should be included in the calculation of the average range. This S, also known as process S, is calculated by the following equation:

$$\text{Process S} = 0.886 \times \text{average range of successive pairs of results of the same mixture.}$$

The process S is particularly suited for concrete (Gibb and Harrison 2010) where there are step changes in average strength in the data set, as the effect of the step change will be limited to a single pair of results. Step change has been discussed in Chapter 7, Basic Statistics. When the process S is substantially different from the traditionally calculated S, it is highly likely that the data set has a step change.

However, there can be a step change even if the two calculations of S do not differ much.

CUSUM Charts

To confirm the step change in 28-day strength, other CUSUM charts such as 7-day strengths, density, temperature, and so forth, are observed. By doing this, a step change can be confirmed as soon as five data points are received (Day 2006).

Use of Control and CUSUM Charts to Analyze Project Test Data

In the following section, both control charts and CUSUM charts have been used to analyze the test results from three different projects supplied by three different producers. These are identified as Project 1, 2, and 3. The raw data are not provided due to space constraints. Table 12.2 shows the averages and other statistical data from the project raw data.

Project 1

The project had almost 300 data points, and the 28-day traditional S is 478 psi (3.3 MPa), which is a Very Good standard of concrete control, according to ACI 214R. The process S based on the average range of consecutive strength test results for this period of data collection is 274 psi (1.89 MPa). As discussed below, the producer had changed the cement content of the picture during the project, which must have contributed to the step change in average strength.

Control charts: Control charts are plotted as Figure 12.1a–g. The 7-day strength test data that were outside the control limits correlated well with 28-day strength test data that fell outside the limits. The 7-to-28-day strength gain is generally within the control limits. One high result (point 205) on the 7-to-28-day strength gain chart coincided with a low 7-day strength but the 28-day result was within control limits. Since this is a single event, it is likely due to problem with the 7-day test rather

TABLE 12.2
Statistical Project Test Data

Property	Project 1	Project 2	Project 3
Average slump, in. (mm)	4.8 (120)	5.2 (130)	7.6 (193)
Average air content,%	2.0	5.6	2.9
Average temperature, °F (°C)	72 (22.2)	68.3 (20.2)	63.1 (17.3)
Average density, lb/ft^3 (kg/m^3)	146.7 (2350)	Not measured	148.8 (2383)
Air-free density, lb/ft^3 (kg/m^3)	149.7 (2398)	Not measured	153.3 (2455)
Average 7-day compressive strength, psi (MPa)	3982 (27.5)	4406 (30.4)	5328 (36.7)
Average 28-day compressive strength, psi (MPa)	5206 (35.9)	5578 (38.5)	9450 (65.2)
Average 7–28 comp. strength gain, psi (MPa)	1224 (8.4)	1173 (8.1)	4123 (28.4)
7-day strength S, psi (MPa)	449 (3.1)	381 (2.6)	961 (6.6)
28-day strength S, psi (MPa)	478 (3.3)	438 (3.0)	1258 (6.6)
Number of data points	293	318	41

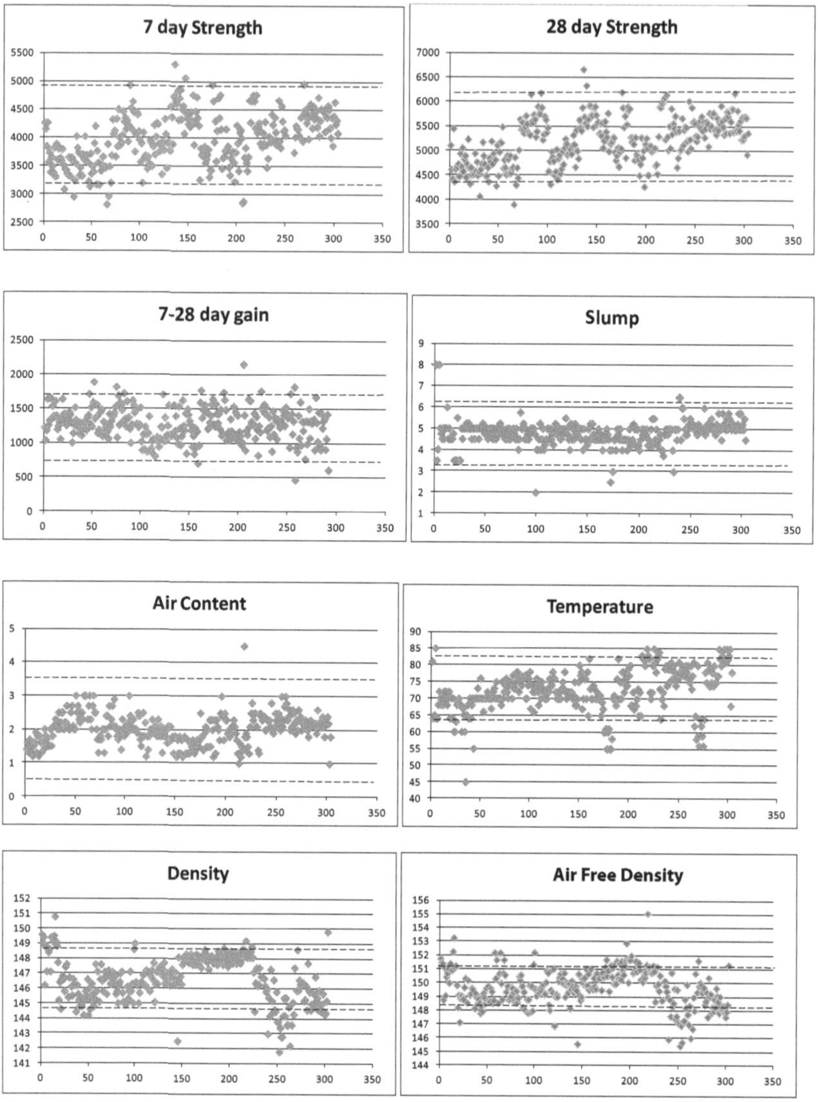

FIGURE 12.1a–g Control charts for Project 1.

than any issue with concrete quality. Other test data from that date showed acceptable strength test results. The 7- and 28-day strength results indicate that the first 70 points or so had lower strengths. Sure enough the concrete producer remarked that cement contents might have been adjusted upward at some point. The slump and air content control charts show very few out-of-control points. The temperature control chart shows a wide difference in concrete temperature, suggesting that set times and possibly water demand might have varied. Unfortunately, slump, air content, density, air-free density, and temperature control charts do not help explain the cause for the out-of-control strength test results.

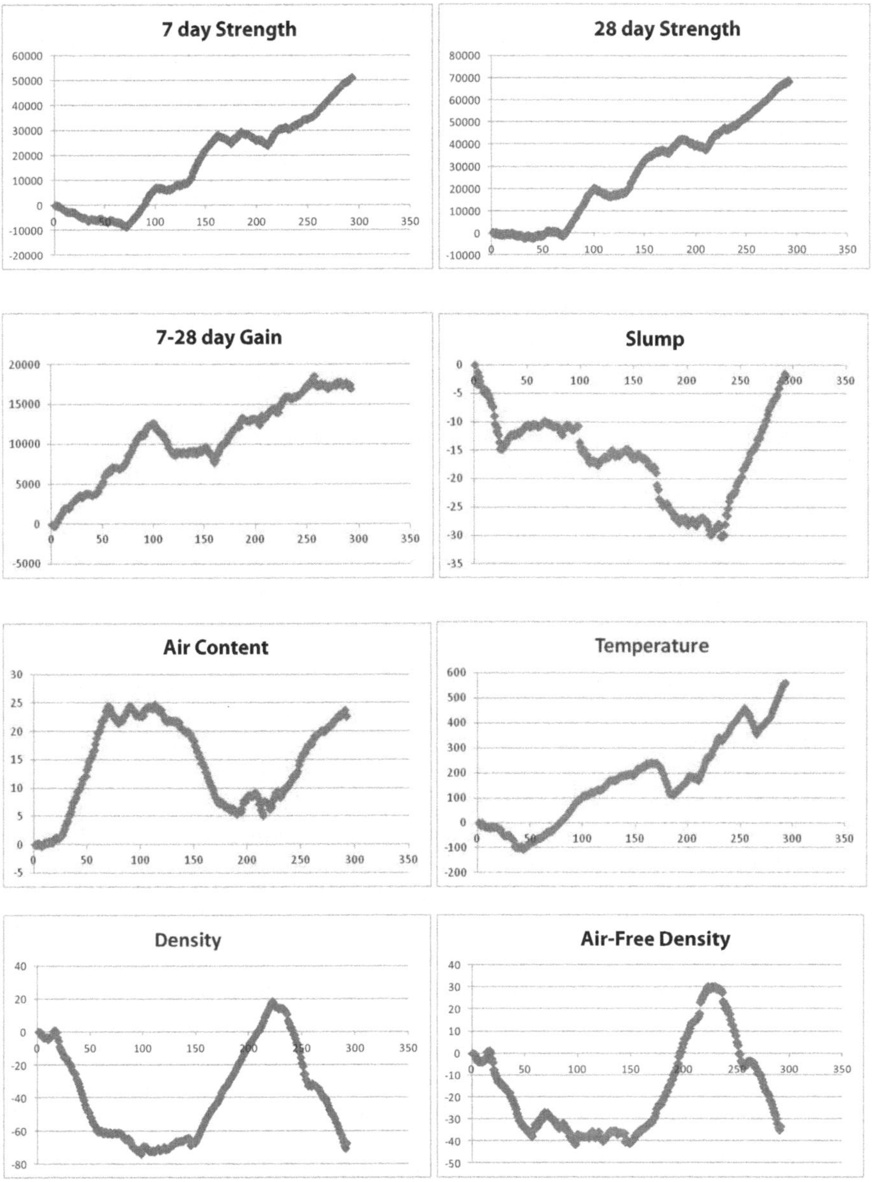

FIGURE 12.2a–g CUSUM charts for Project 1.

CUSUM charts: The CUSUM charts are plotted as Figure 12.2a–g. The 7-day strength CUSUM chart correspond very well with the 28-day strength CUSUM chart. Step changes (changes in average value) in 7- and 28-day strengths are very clearly visible in the CUSUM chart as opposed to the control chart. For the first 70 data points, there is a slight decrease in the 7-day-strength CUSUM charts, but there is negligible step change in the 28-day strength CUSUM charts. After point

70, both the strength CUSUM charts show a positive step change about 300 psi (2.1 MPa). The 7-to-28-day strength gain shows a generally positive step change. Slump and air content CUSUM charts are relatively consistent. However, beyond point 220, there is a clear increase in slump, air content, and temperature, and there is a decrease in density and air-free density. These observations suggest that the mixing-water content and air content have both increased beyond point 220. The step change in air-free density is about –1.2 lb/ft^3 (–19.2 kg/m^3). This level of air-free density change is likely from an increase in mixing-water content of 20 lb/yd^3 (11.9 kg/m^3), assuming that the batched quantities have not changed. From the air content CUSUM chart, the step change can be estimated at about 0.25%. The higher air content and mixing water content should have led to lower strengths beyond point 220. However, that is not observed. Part of the reason for this could be the fact that the concrete producer changed the cement content during the project. CUSUM chart calculations should be started again when mixture proportions are changed.

Project 2

The project had more than 300 data points, and the 28-day strength S is 438 psi (3.02 MPa), which is a "Very Good" standard of concrete control, according to ACI 214R. The process S, based on the average range of all the data points, is 384 psi (2.65 MPa).

Control charts: Control charts are plotted as Figure 12.3a–g. The out-of-control 7-day strength test data do not show a good correlation with the out-of-control 28-day strength test data. This is because the 7-to-28-day strength gain is significantly variable with several points outside the control limits. There are some data points where there is less than a 100 psi (0.7 MPa) 7-to-28-day strength gain! Table 12.3 summarizes some of the out-of-control 7- and 28-day strength test data. It is clear that the 7-to-28-day strength gain was out of control in all of those cases. A closer look shows that very low 7-to-28-day strength gain occurred on certain occasions, suggesting that it might be related to curing/testing practices. For example, six batches of concrete were tested on the same date. Five of those batches had 7-to-28-day strength gain exceeding 1000 psi (6.9 MPa), whereas one of the batches had 7-to-28-day strength gain of only 55 psi, clearly suggesting that the very low strength gain is likely due to poor cylinder fabrication, curing, and testing. On the other hand, on two consecutive days, the average 7-to-28-day strength gain averaged about 400 psi, which is much lower than normal. However, this consistent pattern may indicate performance variation due a specific cement shipment, for example. A close look at the ASTM C917 cement test results from that period is warranted. The slump and air content control charts show very few out-of-control points. Temperature control chart shows a few points on the lower end of the control limits, suggesting that set times and possibly water demand might have varied. The within-batch range shows that, overall, about 8% of all test results had within-batch range greater than 8%, which is higher than the 5% that is considered acceptable according to the ASTM C39 precision statement. The moving 10 test V_1 shows that V_1 exceeded 4% in many of the cases. Both of these findings, as well as the unusual 7-to-28-day strength gain suggest that the testing quality needs to be looked into.

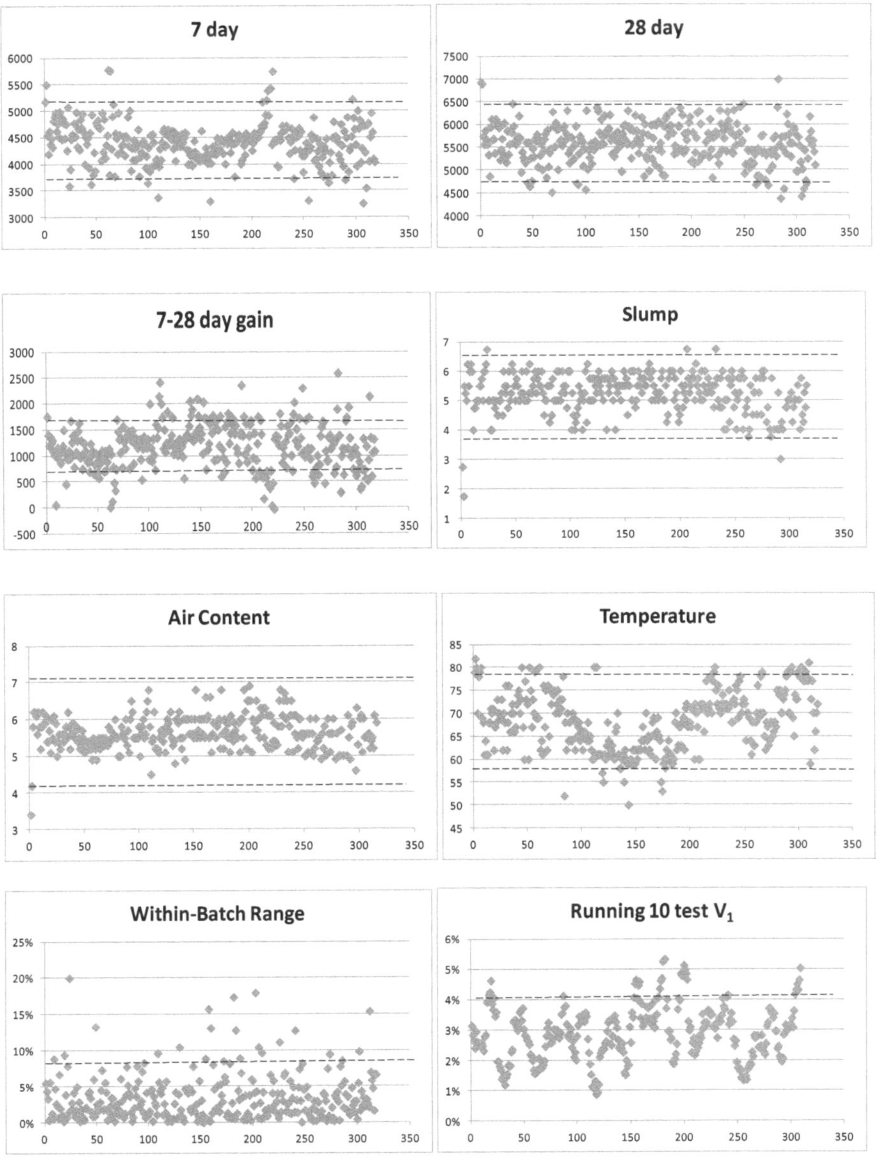

FIGURE 12.3a–g Control charts for Project 2.

CUSUM charts: The CUSUM charts are plotted as Figure 12.4a–g. The 7-day strength CUSUM chart does not always correspond with the 28-day strength CUSUM chart. For the first 100 points, the 28-day CUSUM chart showed an average step change (about –160 psi [–1.1 MPa]), which corresponds well with the 7-day CUSUM chart (about –100 psi [–0.7 MPa]). However, between points 100 and 200, the 28-day CUSUM chart shows a positive step change, whereas the 7-day CUSUM chart shows a negative step change. The difference between 7-day and

TABLE 12.3
Summary of Some of the Out-of-Control Points from Project 2

7-Day Strength, psi (MPa)	28-Day Strength, psi (MPa)	7-to-28-Day Strength Gain, psi (MPa)
3370 (23.2)	5515 (38.0)	2145 (14.8)
3300 (22.8)	4975 (34.3)	1675 (11.6)
3310 (22.8)	5040 (34.8)	1730 (11.9)
4080 (28.1)	4370 (30.1)	290 (2.0)
4020 (27.7)	4470 (30.8)	450 (3.1)
5780 (39.9)	5790 (39.9)	10 (0.1)
5740 (39.6)	5710 (39.4)	−30 (−0.2)

28-day CUSUM charts can be explained by observing the 7-to-28-day strength gain CUSUM chart. The 7–28-day strength gain CUSUM did not show much step change till point 100. However, between points 100 and 200, there is an average positive step change (about 200 psi [1.4 MPa]). So between those points, the decrease in 7-day strength is more than compensated by an increase in 7-to-28-day strength gain. The cumulative effect of the two was an increase in the 28-day strength. The underlying reason for this can be deduced from the concrete temperature CUSUM chart.

For the first 100 data points, the temperature CUSUM chart shows negligible step change. Between points 100 and 200, there is a significant step change (–5°F) (–2.8°C), which was accompanied by a negative step change in 7-day strength, and positive step change in 7-to-28 strength gain and 28-day strength. Between points 200 and 300, there is a significant step change 5°F (2.8°C), which was accompanied by a negative step change in 28-day strength and negative step change in 7-to-28 strength gain. These observations suggest that cylinders might have been cured improperly in the field. Slump and air content appear to show positive step changes, but the actual magnitudes are pretty low (average of 0.25 in. [6 mm], 0.2%, etc.) to influence concrete strength. The within-batch range CUSUM suggests that for the first 150 points there was a negligible step change, but from point 150 onward there was a positive step change, which indicates a change in a testing technician, that is, a trend toward poorer testing quality. The step change of about 0.8% suggests it was not negligible when compared with the overall average within-batch range of 3.3%. The moving 10 test V_1 CUSUM closely matches the within-batch range CUSUM. Overall, the analysis reveals that the concrete is of good quality (slump, air content within control, and strength with low S) with the curing/testing quality varying somewhat.

Project 3

The project had 41 data points and the 28-day S is 1258 psi (8.7 MPa), which is a Poor standard of concrete control, according to ACI 214R. The process S based on the average range of all the data points is 1024 psi (7.06 MPa). In general, the quality control processes of this set of data is likely relatively poor, and establishing assignable causes for the variation will be difficult.

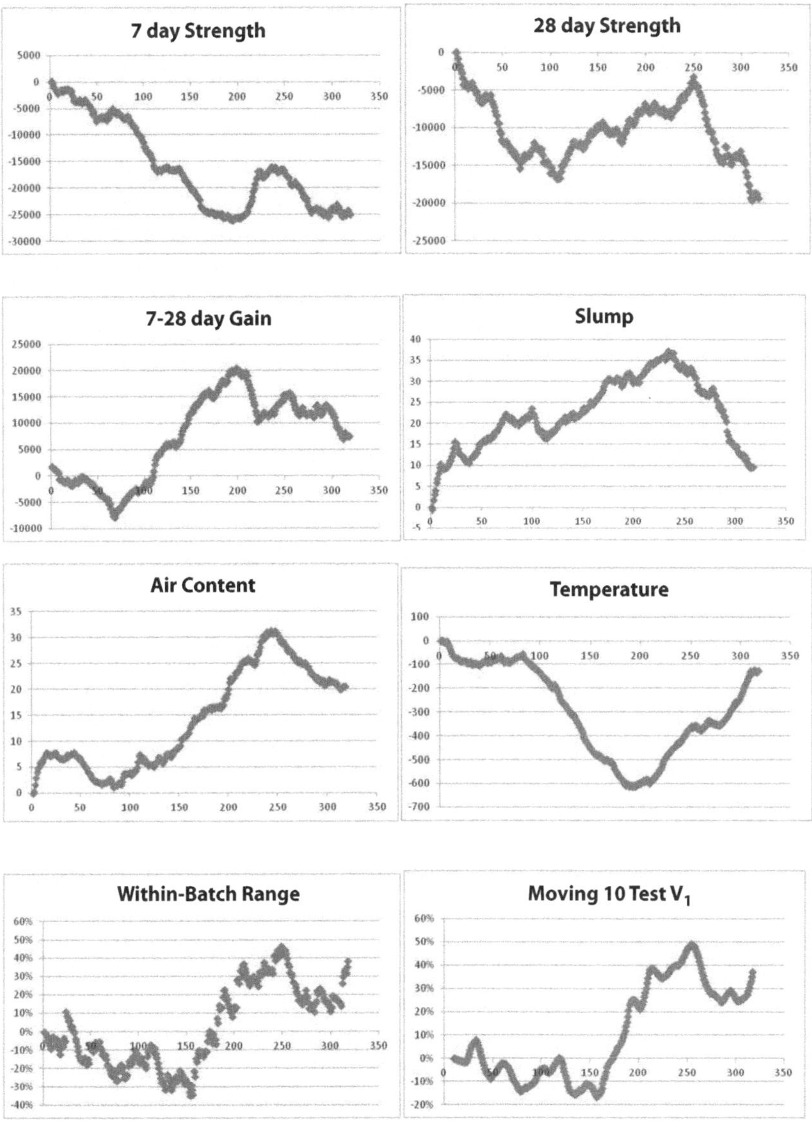

FIGURE 12.4a–g CUSUM charts for Project 2.

Control charts: Control charts are plotted as Figure 12.5a–g. The 7-day strength test data falling outside control limits correlate well with the corresponding 28-day strength test data, as well as the 7-to-28-day strength gain. Points 10 and 20 exceed control limits on the high and low side, respectively. Point 10 had high density and air-free density, suggesting that both mixing-water content and air content maybe low. Point 20 had acceptable density and low air-free density, suggesting that air content was acceptable while mixing water content was high. The corresponding slump for point 20 was also high (10 in.; 255 mm). Several batches also resulted in

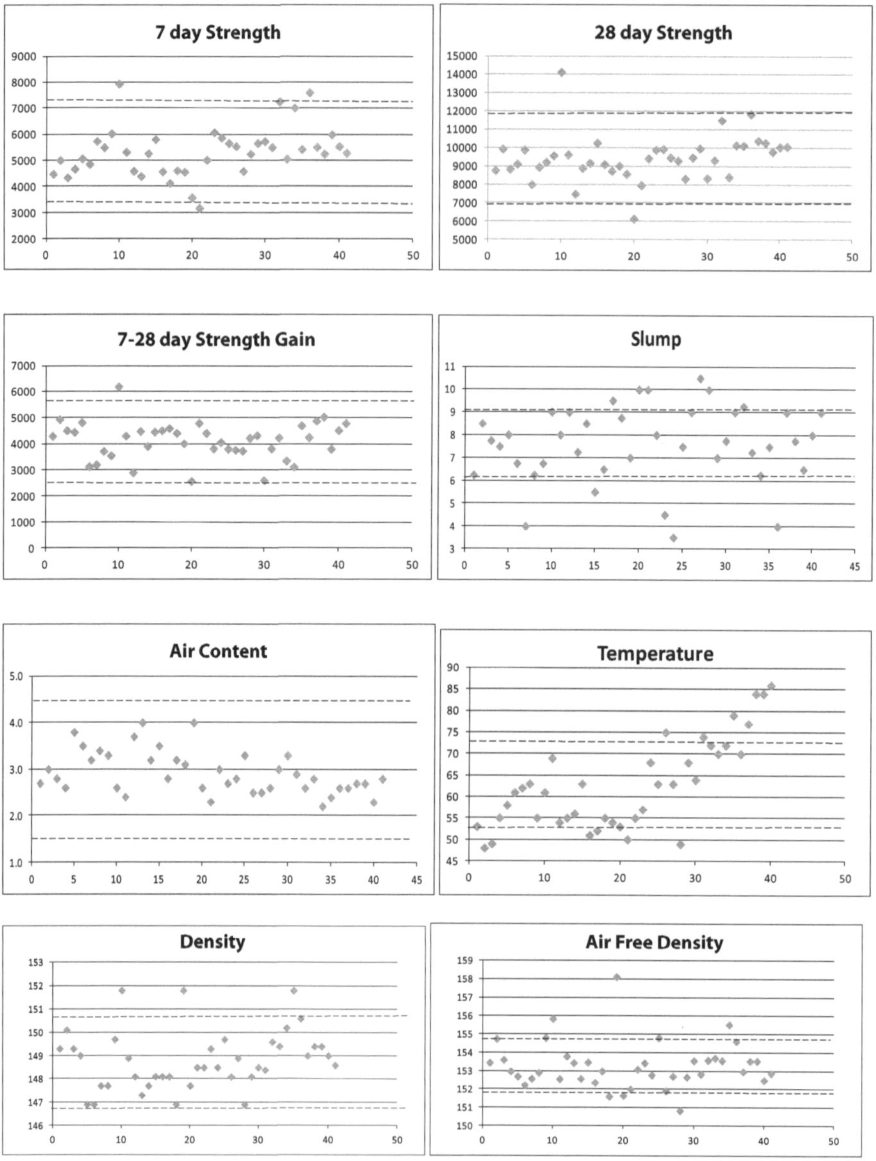

FIGURE 12.5a–g Control charts for Project 3.

measured density and slump being outside the set control limits. However, these batches had acceptable strengths. A temperature control chart shows several points exceeding control limits and a rising trend in the later part of the project, suggesting that set times and possibly water demand might have varied.

CUSUM Charts: The CUSUM charts are plotted as Figure 12.6a–g. The 7-day strength CUSUM chart corresponds well with the 28-day strength CUSUM chart.

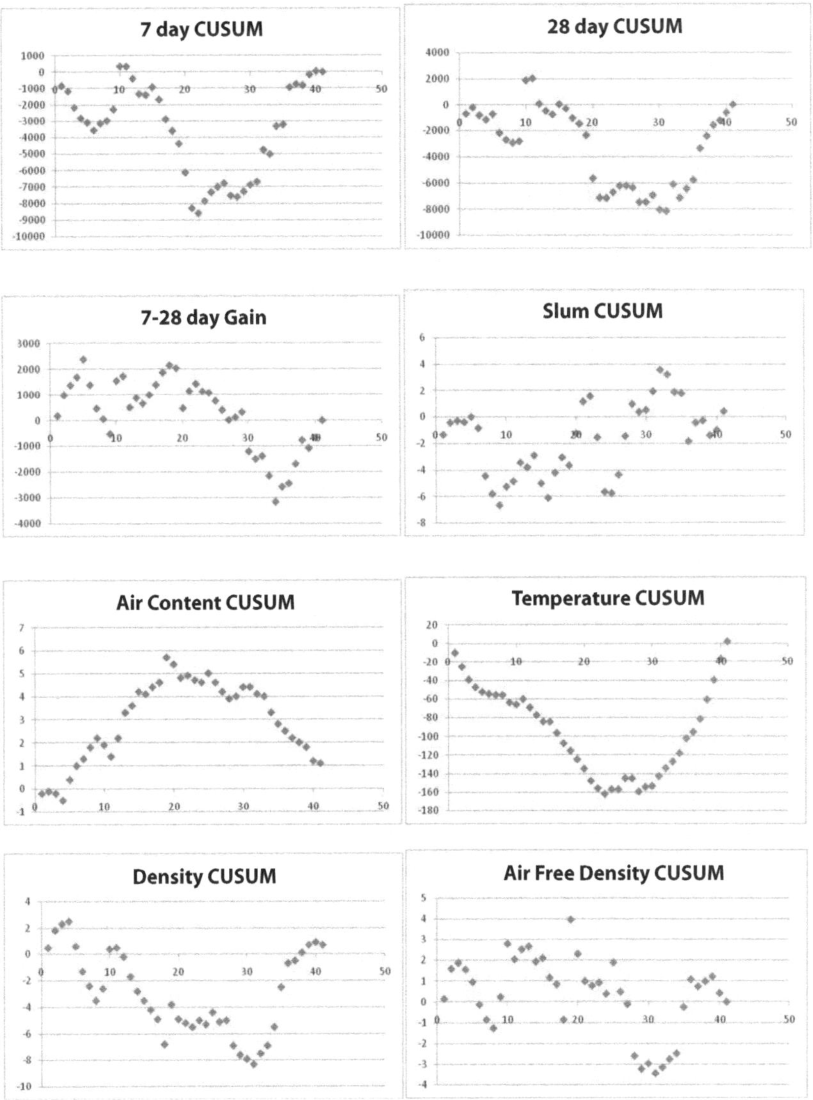

FIGURE 12.6a–g CUSUM charts for Project 3.

Between points 15 and 21, there is a negative step change in strength. Between points 30 and 41 there, is an average positive step change in 7-day strength (about 700 psi) (4.8 MPa), 28-day strength (about 800 psi) (5.5 MPa), and 7-to-28-day strength gain (about 100 psi) (0.7 MPa). Concrete density CUSUM chart shows a negative step change between points 5 and 10, negative step change between 15 and 20, and finally positive step change from point 31 onwards (average about 1 lb/ft^3 [16 kg/m^3] in all cases). The step changes in the 28-day strength CUSUM chart are the exact inverse of that observed in the density CUSUM. Overall, the analysis reveals that the concrete is not of good quality (density varying and strength with high S).

SUMMARY

When concrete test data become available, the results should be entered in a spreadsheet and visualized using control charts and CUSUM charts. A methodology has been provided by which test results from different mixtures are plotted on a single control and CUSUM chart for that property. This ensures that variation due to assignable causes, including step changes (sustained changes in average value), can be identified rapidly. CUSUM charts are more effective than control charts in rapidly identifying step changes, as well as the magnitude.

An investigation should be started as soon as the 28-day (preferably 7 day) strength test results fall outside the control limits or there is a change in slope of the 28-day and 7-day CUSUM chart that indicates a step change. To identify the underlying cause, the investigation should look at control and CUSUM charts of all the properties. At the minimum, control and CUSUM charts on slump, air content, temperature, density, air-free density, compressive strengths (7- and 28-days), 7-to-28-day strength gain, and within-batch 28-day strength range should be developed. Three data sets from actual projects have been provided to illustrate how control and CUSUM charts can be used. It is possible to achieve reduced variability in concrete by monitoring control and CUSUM charts over an extended period (up to 1 year) and by making appropriate changes to quality practices identified as assignable causes in the analysis.

Standard deviation of 28-day compressive strength test results should be monitored based on moving 15 consecutive strength test results. A different way of calculating S has been discussed. When the process S is substantially different from the traditionally calculated S, it is highly likely that the data set has a step change. However, there can be a step change even if the two calculations of S do not differ much. The ultimate aim should be to lower the S. For air-entrained (AE), concrete producers should target traditional S below 450 psi (3 MPa) and 350 psi (2.4 MPa) for non-AE concrete, which corresponds to very good quality control. A very high S can make it difficult to identify special causes or step changes because the overall random variation of the quality process is too high. A very high S leads to a broad range of control limits on the strength control chart, and just because the strength tests fall within this broad range does not indicate that concrete quality is good.

13 Impact of Specifications on Concrete Quality

Concrete specifications can play an important role in ensuring good quality concrete. This chapter makes suggestions on concrete specifications with the aim of improving concrete quality.

ALLOW USE OF STANDARD DEVIATIONS NOT JUST OVER DESIGNS

Almost all concrete project specifications have a specified compressive strength, f'_c, requirement for concrete mixtures. Mixtures are designed to meet the required average compressive strength, f'_{cr}, to ensure that the strength tests meet the strength acceptance criteria. ACI 318 and 301 provide two options to establish the required average strength f'_{cr}. These options are provided in Appendix A. The first option, referred in this chapter as Option A, makes use of the standard deviation, S, established from past strength test records, while the second option, referred in this chapter as Option B, adds a fixed strength increase when no past test records are available.

Many specifications require concrete mixtures to be overdesigned by 1200 psi (8.3 MPa) above the specified strength, that is, the use of Option B. Forcing the use of Option B when a test record is available is not desirable. The standard deviation, S, from a strength test record is a measure of the ability of the concrete supplier to control the quality of the product. It should be noted here that testing is a component of this standard deviation. There should be a benefit and incentive to use this information. Table 13.1 compares the required average strength for a specified strength of 4000 psi (27.6 MPa) by Option A and B for different levels of concrete quality (ACI 214R) as measured by S. Option B requires an f'_{cr} of 5200 psi (35.9 MPa), regardless of the producer's standard deviation. On the other hand, if Option A (based on past test data) is used, a producer with S = 350 psi (2.4 MPa) has to design the mixture for an average strength of 4470 psi (30.8 MPa), whereas a producer with S = 1250 psi (8.6 MPa) has to target at least 6410 psi (44.2 MPa).

For this example, with a specified strength of 4000 psi (27.6 MPa), when the producer's standard deviation is greater than 730 psi (5 MPa), Option B will allow a lower average strength than Option A. A specification that requires Option B does not benefit a quality conscious producer but does benefit the one with higher variability! Option B is surely not a conservative option when S > 730 psi (5 MPa) and can lead to a higher number of strength problems during the project. Statistically, it can be shown that for f'_c = 4000 psi (27.6 MPa) for a concrete with average strength of 5200 psi (35.9 MPa) with S = 1250 psi (8.6 MPa), the probability of failing the strength acceptance criteria is about 10 times higher compared to one with S = 730 psi

TABLE 13.1
Target Average Strengths for f'_c = 4000 psi (27.6 MPa)

QC Standards (ACI 214R)	Excellent	Very Good	Good	Fair	Poor		
S, psi (MPa)	350 (2.4)	450 (3.1)	550 (3.8)	650 (4.5)	750 (5.2)	950 (6.6)	1250 (8.6)
f'_{cr}, psi (MPa)(Option A)	4470 (30.8)	4600 (31.8)	4780 (32.9)	5020 (34.5)	5250 (36.2)	5710 (39.4)	6410 (44.2)
f'_{cr}, psi (MPa) (Option B)				5200 (35.9)			

(5.0 MPa) and the same average strength! In summary, specifications that require Option B merely tell the producer not to pay attention to quality.

MOVE FROM PRESCRIPTIVE TO PERFORMANCE-BASED SPECIFICATIONS

Concrete mixtures designed to prescriptive specifications are typically substantially overdesigned. It is likely that for a given set of materials, a knowledgeable engineer of record can optimize the mixture proportions to meet the performance criteria needed by the specification. However, the engineer cannot specify that optimized mixture proportion in a prescriptive specification. Frequently, a project specification is written for a large geographical area—the whole state in the case of a transportation agency or even the whole country, in the case of some large nationwide design firms. It is impractical to identify an optimized mixture proportion for the broad range of materials that could be encountered. Even if the same set of materials are used, the optimized mixture proportions may not be used in conjunction with lower-quality manufacturing, construction, and testing practices. Clearly, the engineer has to set prescriptive requirements so that the performance criteria are attainable with the lowest level of quality in materials, manufacturing, construction, and testing practices. This is one of the main reasons why prescriptive specifications are substantially overdesigned, frequently with much higher minimum cementitious content requirements than necessary to attain the performance requirements.

MINIMUM CEMENTITIOUS CONTENT

Minimum cementitious (CM) content requirements are commonly used in prescriptive specifications. The ACI 318 Building Code does not have a minimum CM content requirement. The ACI 301 Specification for Structural Concrete has a minimum CM content requirement only for trowelled concrete floors with consideration to achieve adequate finishability. ACI 301 allows for relaxing the minimum CM content requirements if adequate finishing quality and strength is demonstrated. In most concrete specifications, minimum CM content requirements are used as a proxy for a low water-to-cementitious materials ratio (w/cm), which is desired for improved concrete durability. A better way for verifying low w/cm is through specifying an

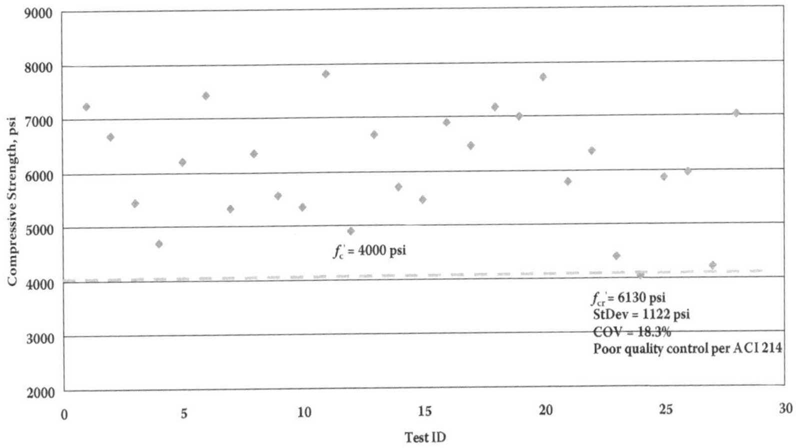

FIGURE 13.1 Variability of compressive strength test results from a project concrete class with minimum cm content requirements.

appropriate compressive strength rather than a minimum CM content. Minimum CM content requirements do not provide any incentive for improving concrete quality, and this becomes obvious from Figure 13.1. In this case, the project had $f'_c = 4000$ psi (27.6 MPa) and a minimum CM content requirement of 650 lb/yd^3 (386 kg/m^3). The test results varied between 4330 psi (29.9 MPa) and 7730 psi (53.3 MPa) with an average of 6130 psi (42.3 MPa), S = 1122 psi (0.8 MPa), resulting in a coefficient of variation of 18.3%. According to ACI 214R, the data suggest that the standard of concrete control was poor. Yet there were no low strength test results and, as a result, there was no incentive to reduce variability and attain improved concrete quality.

Maximum w/cm

It is well understood that concrete permeability reduces with decreasing w/cm. The cementitious composition also impacts permeability. ACI 318 Building Code requires w/cm between 0.40 and 0.50, depending upon specific exposure categories such as freeze thaw, sulfates, or chlorides. Low w/cm concrete has become synonymous with better concrete, and there has been a tendency for engineers to frequently specify low w/cm concrete, even when these exposure conditions do not exist, such as for an interior column. A maximum w/cm specification will typically result in a compressive strength much higher than the specified strength requirement and as pointed out earlier this does not provide any incentive for improving quality. To start with maximum w/cm should be specified only where it is essential. Whenever a maximum w/cm is specified, a higher compressive strength consistent with the w/cm requirement can be specified. The higher strength can be utilized to optimize structural design. Another approach is to move away entirely from a maximum w/cm requirement but instead specify the corresponding durability performance requirement such as a rapid indication of chloride ion penetration (RCP) value (ACI ITG-8R-10) as performed according to the ASTM C1202 test.

CHANGES TO MIXTURE PROPORTIONS AFTER SUBMITTAL

Once a concrete mixture proportion is submitted for a specific class of concrete in a project the producer is held to the same ingredient weights for that class for the duration of the project. The producer is typically allowed to vary only the admixture dosage and attain the concrete performance properties, such as compressive strength, air content, and slump. Large volume ready mixed concrete plants receive multiple shipments of cement and aggregate on a daily basis. Even though the material sources are the same, concrete performance can vary as shipments change. In addition, changing temperatures can result in a change in concrete performance. A specific performance requirement such as a 28-day compressive strength requirement can be consistently attained with varying material shipments and temperatures by designing the mixture for a higher average strength, taking into account the material and temperature variations expected during the project. This is current standard practice. The producer can make use of semi-adiabatic calorimetry, accelerated-cured 2-day cylinder testing, and 7-day standard cured cylinder testing to predict the strengths of cylinders at 28 days. If a lower 28-day strength is expected, the producer can make minor adjustments to the mixture proportions such as lower w/cm. This will result in two benefits: (1) frequency of lower strength test results and resulting expensive investigations will decrease, and (2) producers can now reduce their average strengths since they can now react on a rapid, continual basis for potential low breaks. The lower average strengths will make the mixture more cost effective and sustainable. Some specifications do allow a range of w/cm or batch weights after submittal while some others allow only a reduction in the w/cm by 0.05 or an increase in cementitious content up to 10%. ACI 301-10 allows adjustments to mixture proportions or changes in materials along with supporting documentation. For some engineers, a low 7-day strength test result may be adequate for lowering the w/cm of the concrete mixture.

QUALIFICATIONS

The project specifications should have a quality assurance section (NRMCA Publication 2PE003, 2009) that has the requirements for the producer, installer, and the testing agency.

Producer Qualifications

The NRMCA plant certification established in 1966 provides a system for establishing that production facilities of ready mixed concrete plants are physically capable of furnishing good quality concrete. It reflects, and in many cases exceeds, the requirements of standard specifications for ready mixed concrete, such as ASTM C 94, AASHTO M 157, and the Concrete Plant Standards of the Concrete Plant Manufacturers Bureau, CPMB 100. A licensed professional engineer inspects the concrete plant once in 2 years and specifically looks at material storage and handling, batching equipment, central mixer, ticketing system, and delivery fleet.

Quality personnel with responsibility for concrete mixtures could be required to be certified as an NRMCA Concrete Technologist Level 2, or equivalent. NRMCA also certifies individuals to Level 3 and Level 4 criteria.

NRMCA recently developed a more comprehensive producer quality certification. The goal of this certification is to ensure that the producer is operating an effective QMS for the production of ready mixed concrete that is well established, documented, and implemented, and facilitates continuous improvement. This certification encompasses plant certification but also includes requirements for company personnel, quality objectives, laboratory testing, specification review process, internal audits, concrete mixture development, ingredient materials testing, internal concrete testing, control of mixing water content, record keeping and retention, and customer satisfaction.

Installer and Testing Agency Qualifications

At least one person on the finishing crew should be certified as an ACI Flatwork Finisher, or equivalent. When requested, the installer should furnish a Quality Plan. Further, the chapter on variations due to testing discussed the requirements in the ACI 318 Building Code that can help improve testing quality.

BONUS–PENALTY PROVISIONS

Bonus-penalty provisions in specifications can help improve concrete quality. Some agencies such as the Port Authority of New York/New Jersey and the Virginia Department of Transportation (VDOT) have incorporated bonus-penalty provisions into their projects. The test results from n sub lots (one test result per sub lot) are used to compute (ACI ITG-8R), a quality index (QI) for compressive strength, as

$$QI = \frac{\bar{X} - LSL}{S}$$

where $\bar{X}$ is the average compressive strength, LSL is the lower specification limit or minimum compressive strength, and S is the standard deviation of compressive strength from n sub lots. QI and n are then used to obtain the percent within limits (PWL) that represents the estimate of the percentage of material within the lot that is above the LSL. A lot may be defined as all the concrete in a bridge deck. Sublots and lots can also be defined as given volumes of concrete. Figure 13.2 is a plot developed between the quality index and the PWL based on the reported values in FHWA-SA-96-026. Table 13.2 shows the pay factor adjustment used by VDOT in a project (Sprinkel 2004).

For a project, if LSL = 4000 psi (27.6 MPa), and $\bar{X}$ = 4600 psi (31.7 MPa), S = 1250 psi (8.6 MPa) will result in QI = 0.48, PWL = 67%, and a 23% penalty if Table 13.2 provisions are adopted. Keeping other values the same, changing S from 1250 psi

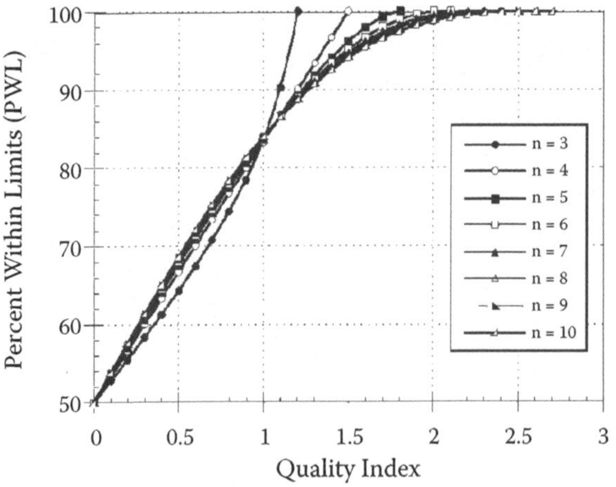

FIGURE 13.2 Relationship between quality index and PWL. (Reprinted with permission from AASHTO R 9, FHWA-SA-96-026, 1996.)

TABLE 13.2
Example of Pay Factor Adjustments to Contract Compensation for Concrete Overlay

Average PWL, %	Pay Factor Adjustment
91 to 100	0.006 (PWL-90)
85 to 90	0.0
55 to 84	–0.9 + 0.01 PWL

Source: Adapted from Sprinkel, M. M., "Final Report, Performance Specification for High Performance Concrete Overlays on Bridges," VTRC 05-R2, Virginia Transportation Research Council, Charlottesville, VA, August 2004, 23 pp.

(8.6 MPa) to 350 psi (2.4 MPa) will result in a QI = 1.71, PWL = 97% and a 4.2% bonus. Clearly, improving concrete quality and attaining a lower S can help transform a substantial penalty into a bonus! Various quality indexes for strength, air content, thickness, smoothness, and rapid indication of chloride ion penetrability (RCPT) are calculated, and the overall quality index is used to calculate the bonus-penalty levels for a project. Bonus-penalty provisions are still new and do have some drawbacks.

Often these penalties are based on the in-place price of the contract, which is significantly more than the price of the concrete. The values of S calculated from a very small number of sub lots can be misleading, which makes it difficult for concrete suppliers and contractors to plan to target a level of the quality index. To protect themselves, they establish very conservative targets and significantly overdesign mixtures and construction processes to avoid penalties. From the contractors viewpoint, increasing the material cost by $4/yd^3 ($5.2/m^3), required for a higher cementitious content, for example, is a justifiable risk to avoid a $400/yd^3 ($520/m^3) penalty on a pay factor for strength tests. In this case, the state highway agency might get what it wants on one particular quality index but might cause another problem on a quality index that may not be part of the payment criteria. For instance, a mixture substantially overdesigned for strength may have a higher propensity to crack.

JOB-SITE CONCRETE ACCEPTANCE TESTING

If there is no quality assurance inspection or testing, there is no incentive to monitor quality. So it is important that a project require acceptance testing for the desired performance. However, it is important to have test criteria that take into account the variability of the different test methods. For example, the testing variability of the ASTM C1202 test method is about four times higher than that of the ASTM C39 strength test. On this basis, it has been suggested (Obla and Lobo 2007, ACI ITG-8R) that the acceptance criteria for the ASTM C1202 should be more lenient than that for C39.

CURRENT INFORMATION ON MATERIAL PROPERTIES

ACI 301 states that aggregates used in the project should conform to ASTM C33, and test results showing conformance should not be older than 90 days, except for test results for soundness, abrasion, and reactivity, which should not be older than 1 year. In addition to the tests required in ASTM C33, tests on relative density, absorption and bulk density of coarse aggregate are required for concrete mixture proportioning and batch weight calculations. It is probably a good idea to use current information (<90 days) on these aggregate properties as they do vary and can result in substantial changes to batch weights and concrete performance. ACI 301 also states that for cementitious materials test results that show conformance to specifications should not be older than 90 days and for admixtures it should not be older than 1 year.

SUMMARY

Concrete specifications should permit the use of the standard deviation from past project test results to establish the average strength f'_{cr} of proposed mixtures for future work. This provides an incentive to producers to target a lower standard deviation. In contrast, defaulting to the fixed strength over the design option does not provide any incentive for improved concrete quality. Moreover, it is not a conservative option when concrete quality is poor and can lead to a higher number of strength problems during the project.

Concrete specifications should reduce prescriptive requirements and incorporate performance requirements. In particular prescriptive minimum cementitious content requirements should be removed. Also, maximum w/cm should not be specified for concrete members that do not require low permeability such as for most interior concrete members. Minimum cementitious content and maximum w/cm requirements typically lead to compressive strengths substantially higher than that required for the specified strength, f'_c, and as a result provide no incentive to target a lower standard deviation and improve concrete quality.

Concrete specifications should have a quality assurance section that incorporates the provided qualification requirements for the company and the personnel involved in the production, installation, and testing of concrete.

Bonus-penalty provisions have adopted some public agencies appear to provide incentive for improving concrete quality through a lower standard deviation. However, these provisions do have some drawbacks.

14 Impact of Concrete Quality on Sustainability

With the increased focus on sustainable construction, building products are being required to document the environmental impact associated with their manufacture and show continuous improvement. For concrete construction, environmental impact can be documented from the source ingredient materials, manufacturing process, and the composition of the concrete mixture. In the use phase of concrete buildings and infrastructure, there are several advantages that can be documented. This article focuses on the concrete mixture and demonstrates how improved concrete quality can play an important role in developing concrete products with reduced environmental impact and contribute to sustainable development.

TARGET A LOW STANDARD DEVIATION

Table 14.1 shows the calculated required average strength for a specified strength of 4000 psi (27.6 MPa) based on the ACI 318 and 301 requirements (Appendix A) for different levels (ACI 214R) of concrete quality as measured by S. Based on the assumption that 1 pound of cement equates to a compressive strength of about 10 psi or 1 kg of cementitious material equates to a compressive strength of 0.15 MPa, Table 14.1 estimates that the cementitious content for the producer with S = 1250 psi (8.6 MPa) is 43% higher than that for the producer with S = 350 psi (2.4 MPa). Table 14.1 also includes the environmental impact of one factor—the carbon footprint—calculated from life cycle inventory data (Marceau et al. [2007]). The CO_2 footprint for the producer with S = 1250 psi (8.6 MPa) is calculated to be about 41% higher than that for the producer with S = 350 psi (2.4 MPa). Using less cement also conserves natural resources associated with its manufacture.

BETTER JOB-SITE CURING AND OVERALL TESTING QUALITY

A component of the strength standard deviation discussed above is associated with acceptance testing of concrete. The factors that can increase the component of variability associated with testing include practices for making specimens; standard curing and subsequent testing–specimen care during initial curing at the job-site; transportation to the lab; curing at the lab; and procedures used to test the specimens for compressive strength. An important effect is temperature and moisture afforded to the test specimens for the initial 24- to 48-hour period after they are cast. Nonstandardized initial job-site curing practices has been shown to lead to more than 1000 psi (7 MPa) reduction in the measured compressive strength test results for a typical 4000 psi (28 MPa) concrete mixture (Obla et al. 2005; Bloem 1969). To

TABLE 14.1
Target Average Strengths for f'_c = 4000 psi (27.6 MPa)

QC Standards (ACI 214R)	Excellent	Very Good	Good	Fair	Poor		
S, psi (MPa)	350 (2.4)	450 (3.1)	550 (3.8)	650 (4.5)	750 (5.2)	950 (6.6)	1250 (8.6)
f'_{cr}, psi (MPa)(Option A)	4470 (30.8)	4600 (31.8)	4780 (32.9)	5020 (34.5)	5250 (36.2)	5710 (39.4)	6410 (44.2)
Cementitious content, lb/yd³ (kg/m³)	447 (265)	460 (273)	478 (284)	502 (298)	525 (311)	571 (339)	641 (380)
CO_2 footprint, lb/yd³ (kg/m³)	463 (275)	476 (283)	494 (293)	518 (307)	541 (321)	587 (348)	657 (390)

avoid expensive investigations, the producer typically compensates by increasing the target average strength of the mixture. This results in much higher cementitious contents and increases the environmental footprint of the concrete mixture. It can be estimated that to overcome the 1000 psi (7 MPa) strength loss due to poor job-site cylinder curing practices the producer would have to increase the cementitious content by 100 lb/yd^3 (59 kg/m^3) or about 20%. During testing, it should be ensured that strength specimens are loaded to complete failure to obtain realistic results and standard deviation of the test results.

MIXTURE OPTIMIZATION

For a given set of materials used in concrete mixtures, strength increases as the w/cm decreases. Mixture proportioning involves establishing a w/cm for the required average strength. However, a specific w/cm can be attained by choosing different levels of cementitious contents and water contents. Table 14.2 derived from an ongoing NRMCA research project provides possible cementitious and water contents for 40% slag cement concrete mixtures at a w/cm of 0.47. It is clear that for a given w/cm the combination of lower mixing-water content and associated cementitious materials

TABLE 14.2
Mixture Proportions for 40% Slag Cement Concrete at Same w/cm

w/cm	CM, lb/yd³ (kg/m³)	Water, lb/yd³ (kg/m³)	Paste, %	Paste/Aggregate Voids	CA/FA, lb/yd³ (kg/m³)
0.47	465 (275)	218 (129)	24	0.94	2163/1396 (1283/828)
	507 (301)	238 (141)	26	1.02	2105/1359 (1249/806)
	528 (313)	248 (147)	27	1.06	2075/1342 (1231/796)
	570 (338)	268 (159)	29	1.14	2020/1305 (1198/774)
	655 (389)	308 (183)	33	1.30	1906/1231 (1131/730)

content leads to a lower paste volume. This paste volume includes air content, which has been assumed to be 2% for these nonair-entrained concrete mixtures. The ratio of paste volume to the combined aggregate void content is also provided. At lower paste volumes, the paste does not completely fill the combined aggregate void content. As the paste volume increases from 24% to 33%, the amount of coarse and fine aggregate decreases by more than 400 lb/yd^3. It was found that as the paste volume increased at the same w/cm the compressive strength was the same, while shrinkage and rapid indication of chloride ion penetrability (ASTM C1202) values increased. A minimum amount of mixing-water content (240–250 lb/yd^3) (142–148 kg/m^3) was required to attain a half-inch (15 mm) slump without admixtures. However, once the mixing water content exceeded the minimum value, higher cement content mixtures tended to further increase the water demand. So it is suggested that for a target w/cm, the lowest amount of mixing-water content for a half-inch (15 mm) slump be used, thereby requiring the lowest amount of cementitious content. Higher workability levels can be attained using water-reducing admixtures. For this set of materials, a mixing-water content of 248 lb/yd^3 (147 kg/m^3) was required to attain a half-inch (15 mm) slump. This corresponds to an optimum cementitious content of 528 lb/yd^3 (148 kg/m^3) for a w/cm of 0.47, thus resulting in a paste volume of 27%. Table 14.2 shows that nonoptimized concrete mixture proportions with paste volume of 33% result in up to a 24% increase in cementitious content.

The NRMCA quality survey of 2012 reported that the target average cementitious content to satisfy the specified strength requirement can be calculated as 452 lb/yd^3 (268 kg/m^3). The 2012 NRMCA SCM survey (Obla et al., 2012) showed that the average nationwide cementitious content used was 561 lb/yd^3 (333 kg/m^3). In other words, the actual cementitious content used was 24% higher than that required to satisfy the strength specification. A major reason is likely the use of prescriptive specifications such as minimum cementitious content requirements, w/cm, and so forth. Optimizing the paste volume for required performance of fresh and hardened concrete is achieved through improved quality management systems, profitability, and environmental impact of concrete mixtures.

FEWER RETURNED CONCRETE AND HARDENED CONCRETE ISSUES

Typically, it is estimated that on average about 5% of concrete is returned to the plant. While some of this is due to over-ordering, about 1%–2% is estimated (Obla 2010) to be concrete rejected for noncompliance with project specifications such as slump, air content, and so forth. If careful attention is paid to concrete quality, the amount of rejected concrete associated with quality problems can be significantly reduced. In some markets, returned concrete can be used to make blocks and can also be used for paving around the plant. Some plants use reclaimers to separate out the aggregates and gray water, while some other plants discharge the returned concrete in an area, allow it to harden, crush it into aggregates, and reuse it. Regardless of the approach adopted for handling returned concrete, it is obvious that reduction in the amount of returned concrete will lead to less energy and resources spent.

Additionally, less concrete will have to be produced and supplied to the project, thereby conserving materials.

Periodically, concrete producers are asked to repair, replace, or mitigate hardened concrete issues because concrete did not meet purchaser's or specification requirements or expectations. Cost to repair hardened concrete can involve core tests (for low cylinder strengths), hiring consultants, evaluating cracking, and so forth, and can become very expensive, even if it does not go to litigation. Removal and replacement can cost 10 times more than the cost of concrete furnished to a project. Additionally, considerable resources are expended to troubleshoot and address hardened concrete issues such as gas spent to visit the job site, new materials consumed in repair, or replacement core testing, which further add to the carbon footprint.

PLANT AND TRUCK MIXER MAINTENANCE

An important aspect of maintaining concrete quality is to maintain concrete plants and trucks in good operating condition and capable of supplying concrete in accordance with ASTM C94. This can be ensured by requiring the NRMCA plant certification program or a suitable equivalent. An aspect of the plant certification program is the annual check of truck mixers for blade wear and concrete buildup. Excessive blade wear and concrete buildup will lead to inefficient mixing, which in turn will lead to poor concrete quality and more fuel consumed for thorough mixing. Excessive buildup of hardened concrete in the range of 5000 lbs can cause more than a 10% increase in energy consumption for mixing and transporting the concrete. In addition, clearly marking the cementitious pipes and having and adhering to a Quality Plan ensure that errors such as pumping cement into fly ash silos can be avoided. This again does impact sustainable production of concrete.

TEMPERATURE MEASUREMENTS

An earlier chapter discussed the importance of tracking concrete and ambient temperatures at the plant with the aim of producing concrete with consistent water content and setting times. Tracking concrete and ambient temperatures can also help one to accurately estimate the amount of heating required for water and aggregate at the plant during cold weather. Similarly, it can help one to accurately estimate the amount of ice needed during hot weather. By preventing overestimation, the energy for heating and cooling the concrete can be reduced.

BATCHING ACCURACY AND YIELD MEASUREMENTS

An earlier chapter discussed that, if producers track the accuracy to which ingredient materials are batched, this can help reduce concrete strength standard deviation and prevent significant material batching errors. The same text discussed how by tracking batching accuracy it was possible to identify early when batching systems are trending to errors and schedule maintenance. Ensuring accurate batching facilitates proper inventory control of ingredient materials, reduced waste, and significant cost savings. In addition, regular measurements of concrete yield can help identify if material batching errors are occurring.

MIXTURE ADJUSTMENTS

As discussed in the previous chapter, by making mixture adjustments based on trends of 7-day strength test results, the producer can avoid potential low strength problems (by day 28) and at the same can also reduce their required average strengths. Both of these will help sustainability.

SUMMARY

Improved concrete quality practices and lower standard deviations of compressive strength result in lower required average strength for a specified strength. This, in turn, will require lower cementitious content in the mixture and hence will result in a lower carbon footprint of the concrete mixture. The carbon footprint for the producer with S = 1250 psi (8.6 MPa) was calculated to be about 41% higher than that for the producer with S = 350 psi (2.4 MPa).

Nonstandardized job-site cylinder curing practices can increase the required cementitious contents for meeting a specified strength by about 20%. Therefore, improved job-site cylinder curing practices can contribute to sustainability.

Concrete mixtures can be optimized to contain low cementitious contents. It is suggested that for a target w/cm the lowest amount of mixing water content for a half-inch slump be used, thereby requiring the lowest amount of cementitious content. Higher workability levels can be attained with admixture dosages.

By paying attention to concrete quality, the amount of rejected concrete, and fresh and hardened concrete issues due to quality reasons can be reduced. This helps sustainability through reduced energy use in handling rejected concrete, supplying new concrete as a replacement, lowering the amount of coring, and lowering the amount of repair materials.

Proper maintenance of plant and mixers ensures reduced energy and waste associated with the production and delivery of concrete. Tracking concrete and ambient temperatures at the plant regularly can help reduce the energy for heating and cooling the concrete. Monitoring batching accuracy ensures conservation of materials while reducing the variability of the concrete produced.

15 Elements of a Quality Management System for a Concrete Producer

Various actions that can help improve quality have been discussed in preceding chapters. A concrete producer needs to put it all together into a Quality Management System (QMS) and adhere to it. A QMS establishes company policy and goals and sets actions and responsibilities for individuals within an organization with regard to quality.

ACI 121R defines a Quality Manual (QM) as the entire body of quality-related documents that collectively describe the QMS. ACI 121R defines a Quality Plan as the overview document that describes the objectives of a QMS. Many construction projects require the submission of a quality plan by the contractor and the concrete supplier.

This article discusses the important elements of a QM and the impact these have on quality. Clearly, if a process does not improve quality and provide a financial or customer benefit it becomes difficult to justify doing that process. This decision can be arrived at only when the impact of a process is measured. It is also important to perform a cost-benefit analysis of a process and to identify variability that cannot be controlled with reasonable cost or resources. NRMCA's Quality Management System for Ready Mixed Concrete Companies (2008) provides detailed guidelines for developing a QM. Other NRMCA publications also address the resources and activities appropriate for a robust quality management system (NRMCA Publication 2P190, NRMCA Quality Control Manual Section 1 and 2). NRMCA recently developed a quality certification program for concrete producers. The certification program sets minimum criteria for a company-established quality manual and audits whether the company does what they state in the QM. The criteria set by the certification program primarily evaluate the impact on the quality of the product from a purchaser's perspective.

WHY SHOULD A COMPANY HAVE A QMS?

A QMS establishes a systematic way of setting quality processes and responsibilities and improving a company's quality. The company's management should be an integral part of establishing and supporting the QMS. An earlier chapter discussed how by improving concrete quality the cost of poor quality can be lowered and the profitability of the company is enhanced. In a Harvard study (Levine and Toffel 2010) of 1000 companies, the companies that adopted and certified to the QMS standard ISO 9001 were compared to those that did not. Citing the Harvard study, Ames et al. (2011) concluded that:

1. Adopters of management system standards have higher rates of corporate survival than nonadopters.
2. Adopters of management system standards have higher sales, employment, payroll, and average annual earnings per employee than nonadopters.
3. Small businesses achieve proportionally more benefits than larger organizations.

For a concrete producer, the QMS also helps with the following:

1. Demonstrate to customers the company's ability to consistently provide ready mixed concrete that meets their performance criteria and/or applicable specifications; this can establish the company as a preferred provider in a market area.
2. Improve customer satisfaction.
3. Provide continuity and uniformity of processes in the various regions in which the company operates, even after key personnel leave the company.
4. Enhance proficiency of personnel and their credibility with customers.
5. Establish a measurement system to facilitate continuous improvement through statistically based concepts, when applicable.

WHAT ARE ELEMENTS OF A QMS AND HOW DOES IT IMPROVE QUALITY?

Quality Objectives and Measurement

An important part of a company's QMS is a statement of quality objectives that are communicated to all personnel. The stated quality objectives in the QM should be measurable and directly related to quality. The measured impact resulting from the quality objectives should also quantify the tangible benefits of improved quality. The following are examples of measurable quality objectives:

1. Maintain concrete strength deviation of the top two selling mixtures for each plant to less than 500 psi. As discussed in Chapter 1 every 100 psi (7 MPa) reduction in the strength standard deviation can help attain a concrete materials cost savings up to \$1.33/yd^3 (\$1.74/m^3).
2. Maintain amount of rejected concrete (as a percent of the concrete produced) for noncompliance with project specifications such as slump, air content, etc., to less than 1%. The benefit can be quantified in terms of revenue generated or saved.
3. Maintain cost to repair, replace, or mitigate hardened concrete issues (cores, etc.) because concrete did not meet purchaser's or specification requirements, expectation, etc., to less than \$0.50/yd^3 (\$0.66/m^3). The benefit is the impact to profitability and optimum use of personnel and resources for proactive quality processes.
4. Maintain customer perception of company's quality through an annual customer survey in which over 90% of responses are Excellent or Very Good

(the choices being Excellent, Very Good, Good, Fair, Poor). The benefit of improved customer perception and resulting repeat business is immeasurable.

5. Average (>10 measurements) relative yield of the top two selling mixtures for each plant between 0.99 and 1.01. Maintaining accurate yield improves customer relations, increases profitability, and reduces performance variations.
6. Number of annual quality-related complaints per 100,000 yd^3 (75,000 m^3) of production less than 4. Benefits result from customer perception and improved use of company resources.

Maintaining on-time delivery, 100% customer satisfaction, use of fuel-efficient trucks, and so forth, though measurable, are not quality related. The same could be said about maintaining plant and personnel certifications, which are seen more as a means to achieving good quality rather than an actual quality objective.

Management Commitment

While most companies may have a written QM, the same cannot be said about the proportion of companies that do what is stated in it. A significant reason for this is management commitment. For a QM to be a living document that sets the standard operating procedures for a company, it is essential that the owner, president, or general manager of the company or division be responsible for approving and signing the QM, and ensuring that it is being implemented. Implementation, monitoring, and making changes to the QM should be the responsibility of the quality committee. The quality manager and some members of the company's senior management should be part of the quality committee. The quality committee should meet at a minimum once every 6 months and discuss changes to the quality practices and the QM. The quality manager in the organization should maintain the QM. Any company employee should be allowed to suggest changes to the QM, and these suggestions should be made to the quality manager. The quality manager works with the operations and sales departments to implement quality initiatives and coordinate all quality activities. So it is beneficial for the company's organizational chart to be structured such that the quality manager reports directly to the general manager of the company or division.

Customer Focus

It is suggested that customer feedback on the concrete quality be sought through an annual survey.

There should be a system for capturing customer complaints and a methodology for addressing them. The methodology should address timelines, responsibilities, and checklists or flowcharts for addressing customer complaints. Complaints should be tabulated for discussion by the quality committee during the internal quality audit. In this manner, customer complaints are used to improve product quality and make changes to the QM.

Customer education seminars of various durations (1 hour to 1 day) should be done periodically. Some topics are scope and qualifications of company's quality organization, basics of quality concrete and proper placement practices, control of mixing-water content, crack prevention, proper curing, performance-based specifications, acceptance test report distribution, proper procedures for sampling, and testing ready mixed concrete. Some presentations can be made to local ACI meetings as well as various engineering groups and contractor associations. The primary aim of these seminars is (1) to show that the company has credible quality personnel and quality practices, (2) to improve specifications to provide incentives for improved quality as discussed in earlier chapters, and (3) to improve job-site concrete handling, curing, sampling and testing procedures to reduce potential liability to the company.

One important forum for customer focus is the prebid, preconstruction, and/or prepour conference. The NRMCA/ASCC Checklist for the Concrete Pre-Construction Conference, developed jointly by NRMCA and the American Society of Concrete Contractors (ASCC) to assist in planning on major and unique concrete projects can be used for this purpose.

Personnel Qualifications

Qualified and knowledgeable personnel are essential for producing quality concrete. The QM should define job qualification requirements, including necessary educational qualifications, experience and scope of responsibilities for quality personnel, plant operators, sales, dispatch, and truck mixer operators. These should be general job descriptions established by the company. This ensures that the company hires qualified personnel on a consistent basis and supports their career growth while employed by the company. The following are suggested personnel qualification requirements:

Quality Manager

The quality manager should have a currently valid NRMCA Concrete Technologist Level 3 Certification or equivalent or should have obtained equivalent training as provided in Appendix E. Seven years experience with specification review and mixture proportioning is desirable.

Plant Operators

Persons in charge of batching concrete should have a current NRMCA Concrete Plant Operator certification or equivalent, or have obtained equivalent training as provided in Appendix E.

Field Testing Technicians

Person(s) conducting field testing should have a current ACI Field Grade I certification or equivalent, which should include at a minimum training and performance evaluation of the following ASTM standards: C31, C138, C143, C172, C173, C231, and C1064.

Laboratory Technicians

Person(s) involved and with responsibility for laboratory tests should have a current ACI Lab Testing Technician Level I or equivalent, which should include at a minimum training and performance evaluation of the following ASTM standards:

C39, C78, C117, C127, C128, C136, C192, C566, C617, C702, C1231, and D75. These represent the basic test methods used to monitor the characteristics of aggregates and of concrete.

Truck Mixer Operators

Truck mixer operators should have a currently valid NRMCA Concrete Delivery Professional (CDP) certification or equivalent or have obtained equivalent training as provided in Appendix E.

Laboratory Testing Capabilities

Companies typically have two types of testing laboratories. There may be a central laboratory that develops the concrete mixtures for the companies. A company may also use a third-party laboratory to perform this type of work. Other laboratories located at the plant might perform routine quality testing, such as performing tests on aggregate grading or moisture content, and various fresh concrete tests, including yield determination. The advantages of both types of laboratories for concrete quality are discussed below.

Central laboratories should meet the requirements of Appendix F at the minimum. If it is an in-house laboratory, it is likely to be serving several concrete plants in the region. Laboratories can seek accreditation by AASHTO, CMEC, A2LA, or other accreditation bodies. Laboratories are advised to participate in the biennial Cement and Concrete Reference Laboratory (CCRL) laboratory inspection program or an alternative third party and make corrective actions to noted items in those inspections. Laboratories should also participate in proficiency sample testing programs, such as those offered by the CCRL or by local groups.

Laboratories located at plants performing routine quality testing need not meet the requirements of Appendix F, but still should maintain proper set of equipment and capabilities to perform the tests listed below. It is recommended that each concrete plant have some testing capabilities with one plant personnel trained to perform the tests.

Company central laboratories should be capable of performing the following tests at a minimum:

Aggregate Tests

Aggregate sampling (ASTM D75), reducing samples to test size (ASTM C702), aggregate moisture content (ASTM C566), sieve analysis of coarse and fine aggregates (C117 and C136), measurement of relative density (specific gravity), and absorption of fine and coarse aggregate (ASTM C127 and C128).

Concrete Tests

Slump (C143), air content (C231, C173), density and yield (C138), temperature (C1064), making and curing cylinders (C192), capping cylinders (C617 or C1231), and compressive strength (C39).

Laboratory facilities that perform routine quality testing should be capable of performing all of the above tests except for ASTM C127, C128, C617, or C1231, C39.

The minimum investment in a central laboratory is the compressive strength testing machine, routine calibration costs, and a constant temperature curing facility. The company needs to employ full-time trained technicians who are overseen by a laboratory manager. The main advantage in having a central laboratory is the capability of developing concrete mixtures that in turn provide opportunity to do mixture optimization. Laboratories can also have advanced testing facilities such as for Alkali Silica Reaction, depending on the market it serves. In this case, the investment in equipment and personnel will be higher, but the capabilities in responding to higher performance-based concrete projects will be greater. The investment will need to be justified to anticipate payback.

Materials Management and Conformance

The company should use ingredient materials that comply with industry specifications, understand the variability associated with each material shipment, and take steps to manage the variability when the materials are used to produce concrete. Concrete ingredient materials should be selected based on performance characteristics and supplier commitment to quality and service and not necessarily on cost and availability. It is not possible for the concrete producer to test each material shipment, and therefore it is important for the producer to ensure that the material supplier follows a good quality program. The following should be considered:

1. Component material variability—Cement strength variation can be evaluated by obtaining ASTM C917 from suppliers; variation in supplementary cementitious materials from supplier reports and variations in aggregate grading or other characteristics that impact the ability to produce consistent concrete.
2. Product meets ASTM requirements, and supplier can provide production variation data at the requested frequency in a timely manner.
3. Whether suppliers have a quality program and QM, an accredited laboratory, and certified quality control technician(s).
4. Performance and compatibility with other ingredients used to produce concrete mixtures.

For each component material, the testing, documentation, and reporting the company requires of its suppliers should be clearly stated in the purchase agreement. Mill test reports of each cementitious material should be received with each shipment or bill of lading. In addition to mill test reports, other cement, fly ash, and aggregate testing to be done by the supplier at the required frequency has been discussed in preceding chapters. All testing results should be reviewed in a timely manner and actions should be taken by the producer. The QM should include testing to be done by the concrete producer and methods of handling nonconforming materials. If the producer actively monitors ingredient material variability, conducts comparative testing, and discusses differences in test results, it will encourage the material supplier to improve their quality practices as well. At the minimum, the concrete producer should conduct moisture content and aggregate grading tests.

Material certifications or statements of compliance by the supplier for all ingredient material sources being used by the company should be collected, reviewed for conformance, and stored. Cements should be in conformance with ASTM C150, C595, or C1157; normal weight aggregates with ASTM C33; lightweight aggregates with ASTM C330; fly ash with ASTM C618; slag cement with ASTM C989; silica fume with ASTM C1240; blended supplementary cementitious materials with ASTM C1697; and chemical admixtures with ASTM C494 and C260. Equivalent AASHTO or other national standards (depending on the operating country) are applicable. When nonpotable sources of water are used, documentation for compliance with ASTM C1602 should be maintained. For chemical admixtures, recommended dosages for various applications and placement conditions should be kept. Chloride contents of each of the admixtures should be known due to restrictions of their use on some projects. Admixture storage should be properly labeled. Admixtures should be protected from freezing or contamination. It is desirable for the dispenser units to be visible from the batcher station for immediate detection of any malfunction. Yard personnel should be trained so that aggregate stockpiles are managed to minimize segregation and breakage, and to ensure consistent characteristics when introduced into mixers. They should be trained on scooping materials to avoid subgrade contamination. Yard personnel should be trained to spot errors in coarse aggregate grading (too coarse or fine), and dirty contaminated aggregates. Lightweight aggregate stockpiles should be sprinkled for saturation. Aggregate moisture content and mixture adjustment should be continuously managed and moisture probe accuracy should be verified on a regular basis. Fine aggregate moistures can be rapidly determined using the Chapman flask (ASTM C70, NRMCA TIP 6 [2011]).

The method(s) used to deliver each component material to the plant(s), the method of storing each material, and how materials are transferred to the mixer should be outlined. There should be a process in place at the plant to verify that material shipments agree with the material order. For example, a shipment identified as C33 No. 57 aggregate should not be accepted if the material order in reality was for a No. 8 aggregate. There should be a process in place so that different cementitious materials are protected from moisture and do not intermingle through the use of separate silos or double bin walls, silo fill pipes are clearly labeled, and locks are provided for fill pipes to ensure the correct material is filled in the silos.

Production Control

The company should clearly understand the variability associated with concrete production and take steps to reduce it. The first step is to ensure that production facilities and delivery vehicles conform to the requirements of ASTM C94. This might be accomplished through the NRMCA certification program or an equivalent inspection and approval process.

Accuracy of scales and volumetric measuring devices should be verified at stated frequencies and procedures established when there is question of measuring accuracy as might be observed in the characteristics of concrete produced. Procedures should be stated to monitor and to address out-of-tolerance batches. Statistical concepts can be used to monitor and improve batching accuracy as discussed in a preceding chapter. Unlike other materials, water can enter a concrete batch from several sources. So the company should follow procedures to control mixing water to within the ASTM C94 tolerance of ±3%. Preventive plant maintenance should be performed as discussed in Appendix G on a regular basis. Trucks should be inspected daily. Procedures should be established to ensure proper mixing of the concrete and for correcting slump and air content to required targets before the truck leaves the plant.

Specification Review, Mixture Development, Optimization

Companies should have a policy and identify responsibilities for reviewing concrete specifications, drawings, and structural notes prior to setting a price for concrete and submitting bids. They should look for the opportunity to discuss the concrete specifications with the architect or engineer (A/E) to promote innovative technology and incentivize concrete quality as discussed in a preceding chapter. There should be a procedure in place to design and submit concrete mixtures that are workable, economical, and meet all of the performance criteria required by the contractor and the design engineer. For most major projects, performance criteria are determined by reviewing the job specifications, drawings, and structural notes. The quality manager should be responsible for identifying all the properties required of fresh and hardened concrete and for developing mixture proportions. There should be a policy for steps to be taken regarding the mixture proportions if material sources change. NRMCA TIP 2 (2011) discusses establishing the required average strength of concrete mixtures. Once the required average strength is established, field or laboratory trial batch data can be used to establish that the proposed concrete mixture proportions will produce an average compressive strength that will equal or exceed the established required average strength. NRMCA TIP 7 (2012) discusses the development of laboratory trial batch data. There may be other prequalification tests for shrinkage, alkali silica reactivity, chloride ion penetrability, or other durability evaluations to be performed. Companies should monitor mixture optimization typically through effective use of cementitious materials. Generally, reducing paste volume can help lower costs and improve performance through reduced shrinkage and thermal effects. Use of a higher percent of SCMs can help reduce costs, but the mixtures need to be designed so that performance requirements such as early age strengths and setting times are still acceptable.

Companies should initiate prebid, preconstruction, and/or prepour conferences to discuss, among other things, issues directly related to concrete quality such as the importance of correct job-site curing and cylinder testing, arranging for concrete supplier to be put on the distribution list of all concrete test reports, and the allowed adjustments to mixture proportions and batch weights.

Receiving Orders and Record Keeping

Companies should define the process for receiving and fulfilling orders for concrete. There should be safeguards to ensure that an order is not skipped or misdirected, and that orders placed much in advance are stored in the system. There should be a procedure to ensure on-time delivery and notifying customers of any delays. Even though much of this is general good business practice, from a quality standpoint, the key is to be able to ensure that when an order is placed the correct concrete mixtures are dispatched to the customer.

Companies should have a policy for record keeping. The process of maintaining batch records, delivery tickets, test data on concrete mixtures performed by the company or by the laboratory performing acceptance testing, material certifications, and material testing by the supplier and the producer should be clearly stated. The period of record retention should be defined in company policy based on type of information, types of project, and jurisdictional requirements.

Testing

Internal Testing at the Plant

The advantages of plant testing, selection of mixture classes, frequency of testing, types of testing, and data analysis and corrective action have been discussed in a preceding chapter. Companies should have a process of incorporating process improvements based on the data analysis. Responsibilities for sampling, testing, data analysis, and corrective procedures should be identified.

Internal Testing at the Job Site

When the company conducts testing to verify the results obtained by commercial testing laboratories, the job-site sampling procedure should be outlined. Ideally, company and commercial lab technicians should perform tests on the same sample of concrete. These data allow a comparison of testing practices, if that is the purpose of internal testing at the job site.

Quality Assurance Test Records

All independent and internal tests performed on the concrete provided by the company should be obtained and recorded for analysis. Control charts and other statistical processes, as outlined in preceding chapters, should be used to identify data trends that could present a problem so that corrective action can be taken before the problem occurs. The company should define the process and personnel responsible for collecting and monitoring quality assurance tests performed by third-party laboratories.

Nonconforming Acceptance Test Results

Company should state procedures and responsibilities for dealing with nonconforming concrete job-site acceptance test results. Personnel that have the responsibility to make adjustments for slump, air content, and water/admixture addition at the job-site should be identified. Water held back (if any) should be noted on the delivery ticket so that water in excess of the design w/cm is not added to the mixture. Job-site added water requested by the purchaser should be recorded and signed off on. Companies should identify personnel who are allowed to make adjustments to mixture proportions and the reasons for the adjustments such as feedback analysis of job-site test data and/or low 7-day strength test results. Plant personnel may have the authority to make minor changes to mixtures such as admixture dosages, and significant modifications to established mixtures, for example, w/cm, would need to be made by the quality manager or his designate.

Procedures for addressing low strength test results and other performance deficiencies should be identified. NRMCA Publication 133 (2011) can be a useful resource for investigating low strength test results. Frequent nonconformances of the same type (low strength, for example) should be investigated, the underlying reason discovered, and changes to the QM made if necessary.

Identification/Traceability

Concrete mixtures provided by the company should have a designation or identification that is understood by company personnel. The quality manager should be responsible for this process and for communicating it internally within the company. Once the project is approved, the appropriate mixture designation should be communicated to the appropriate production personnel who will be responsible for verifying the correct mixtures are produced and dispatched to the customers. Batch records should be identifiable to delivery tickets. Most plants have computerized systems to record actual batch weights. Linked to the batch records should be information that includes concrete truck number, driver name, plant, date/time of batch, project name, location and method of placement, and job-site adjustments, if any. This information is useful when troubleshooting concrete problems in the field and for future data mining for quality enhancements.

Quality Audit

An internal audit is an important process to be conducted by the quality manager to ensure that personnel down the line are performing their delegated quality tasks as defined in the QM and to obtain feedback on any necessary changes in the process for improved efficiency. An internal audit checklist should ideally be developed to maintain a record of quality activities. Internal audit process should be stated in the QM. Internal quality audits should be conducted at least once a year. The audit team assigned to conduct the audit should be identified and records of each audit should be maintained by the quality manager. The results of each audit should be reviewed and discussed at the following quality committee meeting. Based on the results of the audit, corrective actions should be identified and provided to the responsible manager who should subsequently be responsible for correcting any nonconformances within a reasonable period of time.

Returned Concrete and Washwater

Company should have a policy on how they manage returned concrete and washwater. Responsibilities should be identified, and the policy and process communicated to relevant plant personnel.

SUMMARY

A concrete producer who is serious about quality should consistently meet the stated specification and purchaser requirements while attaining concrete with a low variability as measured by the standard deviation of compressive strength test results. The quality management system should be geared toward continuous improvement and positively contribute to the company's bottom line. This objective cannot be achieved without a quality management system with executive management commitment and oversight. In the case of a quality manual, size is not important. It should set reasonable activities and establish a company culture all company personnel always follow. Needless to say, it is not worthwhile to include processes that do not help improve quality and provide financial value.

References

REFERENCED STANDARDS AND REPORTS

American Concrete Institute

1. ACI 117-10, Specifications for Tolerances for Concrete Construction and Materials and Commentary.
2. ACI 121R-08, Guide for Concrete Construction Quality Systems in Conformance with ISO 9001.
3. ACI 211.1-91, Standard Practice for Selecting Proportions for Normal, Heavyweight, and Mass Concrete (Reapproved 2009).
4. ACI 214R-11, Guide to Evaluation of Strength Test Results of Concrete.
5. ACI 221R-96, Guide for Use of Normal Weight and Heavyweight Aggregates in Concrete (Reapproved 2001).
6. ACI 301-10, Specifications for Structural Concrete.
7. ACI 304R-00, Guide for Measuring, Mixing, Transporting, and Placing (Reapproved 2009).
8. ACI 311.6-09, Specification for Ready Mixed Concrete Testing Services.
9. ACI 318-11, Building Code Requirements for Structural Concrete and Commentary.
10. ACI ITG-8R-10, Report on Performance-Based Requirements for Concrete.
11. ACI Concrete Field Testing Technician—Grade I Certification, http://www.concrete.org/Certification/CertificationPrograms.aspx
12. ACI Concrete Flatwork Finisher and Technician

More information about these references can be obtained from

American Concrete Institute
38800 Country Club Drive
Farmington Hills, MI 48331
www.concrete.org

American Society of Testing Materials

1. ASTM C29/C29M-09, Standard Test Method for Bulk Density ("Unit Weight") and Voids in Aggregate.
2. ASTM C31/C31M-12, Standard Practice for Making and Curing Concrete Test Specimens in the Field.
3. ASTM C33/C33M-13, Standard Specification for Concrete Aggregates.
4. ASTM C39/C39M-12a, Standard Test Method for Compressive Strength of Cylindrical Concrete Specimens.
5. ASTM C40/C40M-11, Standard Test Method for Organic Impurities in Fine Aggregates for Concrete.
6. ASTM C42/C42M-13, Standard Test Method for Obtaining and Testing Drilled Cores and Sawed Beams of Concrete.

7. ASTM C78/C78M-10e1, Standard Test Method for Flexural Strength of Concrete (Using Simple Beam with Third-Point Loading).
8. ASTM C87/C87M-10, Standard Test Method for Effect of Organic Impurities in Fine Aggregate on Strength of Mortar.
9. ASTM C88-13, Standard Test Method for Soundness of Aggregates by Use of Sodium Sulfate or Magnesium Sulfate.
10. ASTM C94/C94M-13, Standard Specification for Ready Mixed Concrete.
11. ASTM C109/C109M, Standard Test Method for Compressive Strength of Hydraulic Cement Mortars (Using 2-in or 50-mm Cube Specimens).
12. ASTM C117-13, Standard Test Method for Materials Finer than 75-μm (No. 200) Sieve in Mineral Aggregates by Washing.
13. ASTM C123/C123M-12, Standard Test Method for Lightweight Particles in Aggregate.
14. ASTM C127-12, Standard Test Method for Density, Relative Density (Specific Gravity), and Absorption of Coarse Aggregate.
15. ASTM C128-12, Standard Test Method for Density, Relative Density (Specific Gravity), and Absorption of Fine Aggregate.
16. ASTM C131-06, Standard Test Method for Resistance to Degradation of Small-Size Coarse Aggregate by Abrasion and Impact in the Los Angeles Machine.
17. ASTM C136-06, Standard Test Method for Sieve Analysis of Fine and Coarse Aggregates.
18. ASTM C138/C138M-13, Standard Test Method for Density (Unit Weight), Yield, and Air Content (Gravimetric) of Concrete.
19. ASTM C142/C142M-10, Standard Test Method for Clay Lumps and Friable Particles in Aggregates.
20. ASTM C143/C143M-12, Standard Test Method for Slump of Hydraulic-Cement Concrete.
21. ASTM C150/C150M-12, Standard Specification for Portland Cement.
22. ASTM C172/C172M-10, Standard Practice for Sampling Freshly Mixed Concrete.
23. ASTM C173/C173M-12, Standard Test Method for Air Content of Freshly Mixed Concrete by the Volumetric Method.
24. ASTM C192/C192M-13, Standard Practice for Making and Curing Concrete Test Specimens in the Laboratory.
25. ASTM C204-11, Standard Test Methods for Fineness of Hydraulic Cement by Air-Permeability Apparatus.
26. ASTM C231/C231M-10, Standard Test Method for Air Content of Freshly Mixed Concrete by the Pressure Method.
27. ASTM C295/C295M-12, Standard Guide for Petrographic Examination of Aggregates for Concrete.
28. ASTM C311/C311M-13, Standard Test Methods for Sampling and Testing Fly Ash or Natural Pozzolans for Use in Portland-Cement Concrete.
29. ASTM C403/C403M-08, Standard Test Method for Time of Setting of Concrete Mixtures by Penetration Resistance.
30. ASTM C494/C494M-13, Standard Specification for Chemical Admixtures for Concrete.
31. ASTM C535-12, Standard Test Method for Resistance to Degradation of Large-Size Coarse Aggregate by Abrasion and Impact in the Los Angeles Machine.
32. ASTM C566-13, Standard Test Method for Total Evaporable Moisture Content of Aggregate by Drying.
33. ASTM C617/C617M-12, Standard Practice for Capping Cylindrical Concrete Specimens.
34. ASTM C618-12, Standard Specification for Coal Fly Ash and Raw or Calcined Natural Pozzolan for Use in Concrete.
35. ASTM C702/C702M-11, Standard Practice for Reducing Samples of Aggregate to Testing Size.

36. ASTM C 917-05, Standard Test Method for Evaluation of Cement Strength Uniformity From a Single Source.
37. ASTM C1064/C1064M-12, Standard Test Method for Temperature of Freshly Mixed Hydraulic-Cement Concrete.
38. ASTM C1077-13, Standard Practice for Agencies Testing Concrete and Concrete Aggregates for Use in Construction and Criteria for Testing Agency Evaluation.
39. ASTM C1157/C1157M-11, Standard Performance Specification for Hydraulic Cement.
40. ASTM C1202-12, Standard Test Method for Electrical Indication of Concrete's Ability to Resist Chloride Ion Penetration.
41. ASTM C1231/C1231M-12, Standard Practice for Use of Unbonded Caps in Determination of Compressive Strength of Hardened Concrete Cylinders.
42. ASTM C1252-06, Standard Test Methods for Uncompacted Void Content of Fine Aggregate (as Influenced by Particle Shape, Surface Texture, and Grading).
43. ASTM C1260-07, Standard Test Method for Potential Alkali Reactivity of Aggregates (Mortar-Bar Method).
44. ASTM C1293-08b, Standard Test Method for Determination of Length Change of Concrete Due to Alkali-Silica Reaction.
45. ASTM D75/D75M-13, Standard Practice for Sampling Aggregates.
46. ASTM D422-63 (2007), Standard Test Method for Particle-Size Analysis of Soils.
47. ASTM D2419-09, Standard Test Method for Sand Equivalent Value of Soils and Fine Aggregate.
48. ASTM D4791-10, Standard Test Method for Flat Particles, Elongated Particles, or Flat and Elongated Particles in Coarse Aggregate.
49. ASTM E178-08, Standard Practice for Dealing With Outlying Observations.
50. ASTM STP 169D, "Significance of Tests and Properties of Concrete and Concrete-Making Materials, edited by Lamond, J.F., and Pielert, J.H., 2006, 655 pgs.

More information about these references can be obtained from

American Society of Testing Materials
100 Barr Harbor Drive
West Conshohocken, PA 19428
www.astm.org

National Ready Mixed Concrete Association

1. Ready Mixed Concrete Industry Benchmarking and Statistics, http://www.nrmca.org/aboutconcrete/surveys.asp.
2. NRMCA Plant and Truck Certification Program, http://www.nrmca.org/Research_Engineering/Plant_Certification/Main.htm.
3. NRMCA CIP 26, Jobsite Addition of Water, Concrete in Practice series.
4. NRMCA Concrete Technologist Certification Instruction Manual, Outline and Tables for Proportioning Normal Weight Concrete—Advanced Version.
5. NRMCA Draft Standard Practice for Evaluation of Uniformity of Fly Ash from a Single Source, 1988, 9 pgs.
6. NRMCA Quality Control Manual Section 3, Checklist for Certification of Ready Mixed Concrete Production Facilities, Eleventh Revision, 2011.
7. NRMCA Plant Inspector's Guide for NRMCA Certification of Ready Mixed Concrete Production Facilities in accordance with NRMCA Quality Control Manual—Section 3, 2011.

8. NRMCA/ASCC Checklist for the Concrete Pre-Construction Conference, 24 pgs.
9. NRMCA Publication 133, In-Place Concrete Strength Evaluation—A Recommended Practice, 2011, 10 pgs.
10. NRMCA Publication 2PE003, Guide to Improving Specifications for Ready Mixed Concrete, 2009, 22 pgs.
11. NRMCA Plant, Personnel, and Product Certifications, http://www.nrmca.org/certifications/.
12. NRMCA Quality Management System for Ready Mixed Concrete Companies, 2008, http://www.nrmca.org/p2p/QMS%203%20parts%20small.pdf.
13. NRMCA Publication 2P190, Guideline Manual for Quality Assurance/Quality Control, 2006.
14. NRMCA Quality Control Manual Section 1, Ready Mixed Concrete Quality Control Guide, 1996.
15. NRMCA Quality Control Manual Section 2, Ready Mixed Concrete Quality Control Check List, 1996.
16. NRMCA TIP 2, Establishing the Required Average Strength of Concrete, Technology in Practice series, 2011, 4 pgs.
17. NRMCA TIP 7, Creating and Using Three Point Curves for Laboratory Trial Batches, Technology in Practice series, 2012, 4 pgs.
18. NRMCA TIP 8, Concrete Yield, Technology in Practice series, 2012, 5 pp.
19. NRMCA CIP 8, Discrepancies in Yield, Concrete in Practice series, 2000.
20. NRMCA Quality Survey, 2012, Concrete InFocus, Winter 2013, Vol. 13, No. 1, pp. OC5-OC10.

More information about these references can be obtained from

National Ready Mixed Concrete Association
900 Spring Street
Silver Spring, MD 20910
www.nrmca.org

Further Reading

AASHTO Accreditation Program (AAP) for Testing Laboratories, http://www.amrl.net/amrlsitefinity/default.aspx.

AASHTO M157, Standard Specification for Ready Mixed Concrete, American Association of State Highway and Transportation Officials, Washington DC., 2013.

AASHTO T 330, "Test Method for the Qualitative Detection of Harmful Clay of the Smectite Group in Aggregate Using Methylene Blue," AASHTO Standard Specifications, Part 2B: Tests, American Association of State Highway and Transportation Officials, Washington, DC., 2000.

Abel, J. D., and Hover, K. C., "Field Study of the Setting Behavior of Fresh Concrete," *Cement, Concrete and Aggregates*, Vol. 22, No. 2, 2000, pp. 95–102.

American Association for Laboratory Accreditation (A2LA) for Construction Materials Testing (CMT), http://www.a2la.org/appsweb/cmt.cfm.

Ames, M., Blake, R., Caruso, M. J., and Heinle, P., "Why Management System Standards Add Value, Part 1 and 2," *Quality Digest*, http://www.qualitydigest.com/inside/quality-insider-column/why-management-system-standards-add-value-part-1.html (accessed December 11, 2012).

AS 1379, "Specifications and Supply of Concrete," Standards Australia, 1997, 38 pgs.

Bain, D., and Obla, K. H., "Concrete Quality Control—The Untapped Profit Center," *Concrete InFocus*, Fall 2007, Vol. 6, No. 3, NRMCA, pp. 63–69.

Bloem, D. L., "Effect of Curing Condition on Compressive Strength of Concrete Test Specimens," NRMCA Publication No. 53, NRMCA, Silver Spring, October 1969, 15 pp.

Bloem, D. L., and Gaynor, R. D., "Effect of properties on strength of concrete, *Journal of the American Concrete Institute*, Vol. 60, No. 10, October 1963, pp. 1429–1456.

Bloem, D. L., and Gaynor, R. D., "Factors Affecting the Homogeneity of Ready-Mixed Concrete," Report No. 1, Phase I, NRMCA, Silver Spring, MD, 1970, 25 pp.

Bury, M. A., Bury, J. R., and Martin, D., "Testing Effects of New Admixtures on Concrete Finishing," *Concrete International*, January 1994.

CCRL Laboratory Inspection Program, http://www.ccrl.us/Lip/LipProgramDescriptions.htm.

CMEC Accreditation for Testing Laboratories, http://cmecweb.org/accreditation/.

Concrete Plant Standards, Concrete Plants Manufacturers Bureau, CPMB 100, 15th Revision, 2007, http://www.cpmb.org/cpmb%20standards/CPMB%20100-07%20reduced.pdf.

Cost, T. "Thermal Measurements of Hydrating Concrete Mixtures—A Useful Quality Control Tool for Concrete Producers," Reviewed and Approved by NRMCA, NRMCA Publication 2PE004, August 2009, 10 pp, www.nrmca.org.

Crammatte, A. B., "Process Control at a Hawaiian Batch Plant," *Concrete International*, Vol. 20, No. 9, September 1998, pp. 41–44, American Concrete Institute, www.concrete.org.

Crandall, J. R., and Blaine, R. L., Statistical evaluation of interlaboratory cement tests," *ASTM Proceedings*, Vol. 59, 1959, pp. 1129–1154.

Daniel, G., "Batch Record Data Compilation and Summary," submitted to the RMC Research and Education Foundation, 2006.

Daniel, G. D., and Lobo, C. L., "User's Guide to ASTM Specification C94 on Read-Mixed Concrete," Co-published by ASTM and NRMCA, 2005, www.nrmca.org, 130 pp.

Davis, R. E., Jr., "Uniformity of Portland Cements—Facts and Fantasies," Presented at the 45th Annual NRMCA and 59th Annual NSGA Convention, New Orleans, LA, January 1975.

Day, K. W., "Role of Specification, Mix Design, QC and QA in Developing Durable, Economical and Sustainable Concrete Construction," Concrete for 21st Century Construction, UKIERI Concrete Congress, New Delhi, March 2011, http://www.kenday.id.au/indexnc.html.

Day, K. W., 2006, *Concrete Mix Design, Quality Control and Specification*, 3rd edition, Taylor & Francis, New York, NY, 357 pp.

Detwiler, R. J., Thomas, W., Stangebye, T., and Urahn, M., "Variability of 4 × 8 in. Cylinder Tests," *Concrete International*, Vol. 31, No. 5, May 2009, New York, NY, pp. 57–61.

EN 206–1, "Concrete—Part 1: Specification, performance, production and conformity," European Committee for Standardization, 2000, 72 pgs.

FHWA-SA-96-026, Quality Assurance Software for the Personal Computer, Federal Highway Administration, Washington DC, 1996, 185 pgs.

Floor Discussion of Paper on "Variations in Portland cement," *Proceedings, ASTM International*, West Conshohocken, PA, Vol. 58, 1958, pp. 1033–1043.

Gaynor R. D., and Meininger, R. C., "Evaluating Concrete Sands: Five Tests to Determine Quality," *Concrete International*, Vol. 5, No. 12, December 1983, pp. 53–60.

Gaynor, R. D., "Avoiding Uniformity Problems in Truck-Mixed Concrete," *Concrete Producer/Concrete Journal Magazine*, 1996, 1975, 24 pp.

Gaynor, R. D., "Cement Strength and Concrete Strength—An Apparition or a Dichotomy," *Cement, Concrete, and Aggregates*, CCAGDP, Vol. 15, No. 2, Winter 1993, pp. 135–144.

Gaynor, R. D., "Cement Strength Variability and Trends in Cement Specifications in the U.S.," Eighth Congress of the European Ready Mixed Concrete Organization (ERMCO), Rome, Italy, May 1986, 8 pp.

Gaynor, R. D., and Mullarky, J. I., "Mixing Concrete in a Truck Mixer," NRMCA Publication No. 148, NRMCA, Silver Spring, MD, 1975, 24 pp.

Gaynor, R. D., Meininger, R. C., and Khan, T. S., "Effect of temperature and delivery time on concrete proportions," *Temperature Effects on Concrete*, ASTM STP 858, T. R. Naik, Ed., American Society of Testing Materials, Philadelphia, June 1985, pp. 68–87, www.astm.org.

Gibb, I., and Harrison, T., "Use of Control Charts in the Production of Concrete," 2010, ERMCO, 50 pp.

Goeb, E. O., "A review of the NMRCAA comparison test program," Conference on Uniformity in the Initial Curing of Concrete Test Specimens, Albuquerque, NM, August 1995.

Goeb, E. O., "Enforcing Standards for Strength Testing," *Concrete Products*, February 1995.

ISO 9001, "Quality Management Systems—Requirements," International Organization for Standardization, 2008, http://www.iso.org/iso/catalogue_detail?csnumber=46486.

Juran, J. M., and De Feo, J. A., *Juran's Quality Handbook—The Complete Guide to Performance Excellence*, 6th edition, McGraw Hill, New York, NY, 2010, 1113 pp.

Klieger, P. L., "Description of NRMCA-PCA Cement Strength Uniformity Test Program," Presented at the NRMCA Opening Session, Annual Convention, Houston, TX, February 1976.

Kosmatka, S. H., Kerhoff, B., and Panarese, W. C., *Design and Control of Concrete Mixtures*, 14th edition, Portland Cement Association, 2002, www.cement.org, 358 pp.

Lane, D. S. "Testing Fly Ash in Mortars for Air Entrainment Characteristics," *Cement, Concrete, and Aggregates*, CCAGDP, Vol. 13, No. 1, Summer 1991, pp. 25–31.

Levine, D. I., and Toffel, M. W., "Quality management and job quality: How the ISO 9001 standard for quality management systems affects employees and employers," *Management Science*, Vol. 56, No. 6, June 2010, pp. 978–996.

Lobo, C. L., "Report on the Round Robin Strength Testing Program-2011," Cosponsored by WACEL and RMC-PTC, June 2011, NRMCA, 9 pp.

Malisch, W. R., Comments made at NRMCA RES Committee Meeting, October 2010, Charlotte, NC.

Marceau, M. L., Nisbet, M. A., and VanGeem, M. G., "Life Cycle Inventory of Portland Cement Concrete," SN3011, Portland Cement Association, Skokie, IL, 2007, 113 pp., www.cement.org.

Mather, B., Interview in ASTM Standardization News, *ASTM International*, July 1975.

Meininger, R. C. "Use of Fly Ash in Air-Entrained Concrete—Report of Recent NSGA-NRMCA Research Laboratory Studies," National Ready Mixed Concrete Association, Silver Spring, MD, February 1981, 32 pp.

Meininger, R. C., "Effects of Initial Field Curing on Standard 28-Day Cylinder Strengths," *Cement, Concrete, and Aggregates*, Vol. 5, No. 2, Winter 1983, pp. 137–141.

Obla, K., Rodriguez, F., and Ben-Barka, S., "Effects on Non-Standard Curing on Strength of Concrete-A Research Project at the NRMCA Research Laboratory," *Concrete Infocus*, Winter 2005, Vol. 3, No. 4, NRMCA, pp. 57–59.

Obla, K. H., "2010 NRMCA Quality Benchmarking Survey," *Concrete InFocus*, Vol. 10, No. 1, January–February 2011, NRMCA, pp. 11–15.

Obla, K. H., and Lobo, C. L., "Acceptance Criteria for Durability Tests," *Concrete International*, Vol. 29, No. 5, May 2007, pp. 43–48.

Obla, K. H., Lobo, C. L., and Kim, H. "The 2012 NRMCA Supplementary Cementitious Materials Use Survey," Fall 2012, www.nrmca.org.

Obla, K. H., Upadhyaya, S., Goulias, D., Schindler, A. K., and Carino, N. J., "New Technology-Based Approach to Advance Higher Volume Fly Ash Concrete with Acceptable Performance," August 2008, 217 pp., www.nrmca.org.

Ozyildirim, C., "Efforts to improve quality assurance and quality control of concrete in Virginia," *Concrete Materials, Transportation Research Record: Journal of the Transportation Research Board*, No. 2240, Washington, DC, 2011, pp. 36–39.

Peters, D. J., "Evaluating Cement Variability—The First Step," A Discussion of the New ASTM Method C917-80 and the Past, Present, and Future for Cement Uniformity Evaluation, NRMCA Publication No. 161, NRMCA, Silver Spring, MD, April 1980.

Richardson, D. N., "Review of variables that influence measured concrete compressive strength," *Journal of Materials in Civil Engineering*, Vol. 3, No. 2, May 1991, pp. 95–112.

Scott, B. M., "Cement Strength Uniformity—A Ready-Mix Producer's View," NRMCA Publication No. 165, NRMCA, Silver Spring, MD, December 1981.

Snell, L. M., "Assessing the Concrete Technicians," *Concrete International*, Vol. 17, No. 3, March 1995, pp. 42–44.

Sprinkel, M. M., "Final Report, Performance Specification for High Performance Concrete Overlays On Bridges," VTRC 05-R2, Virginia Transportation Research Council, Charlottesville, VA, August 2004, 23 pp.

Suprenant, B. A., and Malisch, W., "Diagnosing Slab Delaminations: Part 3," *Concrete Construction*, March 1998, pp. 277–283, www.hanleywood.com.

Walker, S., and Bloem, D. L., "Variations in Portland cement—part 2," *Proceedings, ASTM International*, West Conshohocken, PA, Vol. 61, 1961, pp. 1059–1077.

Walker, S., and Bloem, D. L., 1958, "Variations in Portland cement," *Proceedings, ASTM International*, West Conshohocken, PA, Vol. 58, 1958, pp. 1009–1032.

Weaver, W. S., Isabelle, H. L., and Williamson, F., "A study of cement and concrete correlation," *Journal of Testing and Evaluation, ASTM International*, West Conshohocken, PA, Vol. 2, No. 4, July 1974, pp. 260–303.

Wills, M. H., Jr., "How aggregate particle shape influences concrete mixing water requirement and strength," *Journal of Materials*, Published by ASTM, Vol. 2, No. 4, December 1967, pp. 843–865.

Glossary

Average of a list of "n" numbers is the sum of the numbers divided by n.

Batch is the quantity of material mixed at one time or in one continuous process. A truckload of concrete is considered a batch.

Coefficient of variation is the standard deviation expressed as a percentage of the average. It is denoted as V.

Composite sample is obtained when two or more grab samples obtained at fixed time or other intervals are mixed together. The variability of test results performed on grab or composite samples can be different.

Control chart is a graphical tool use to determine whether a process is in control over time. Control charts are discussed in depth in the chapter on Basic Statistics.

CUSUM chart is a type of control chart used to detect step changes, that is, change in mean value. CUSUM charts are discussed in depth in the chapter on Basic Statistics.

Grab sample is a random sample obtained in one operation from a lot.

Population includes all the possible items that characterize a physical property of a class of objects or materials. It includes all the parts of the total. *Examples:* Thickness of the entire pavement, strength of concrete in a column.

Quality assurance (QA) is defined by ACI 121R-08 as actions taken by an organization to provide and document assurance that what is being done and what is being provided are in accordance with the contract documents and standards of good practice for the work. According to Juran, the main purpose of quality assurance is to verify that control is being maintained. Performance is evaluated after operations, and the resulting metrics are used to determine conformance to customer needs and expectations.

Quality control (QC), also called *process control*, is defined as those actions and considerations necessary to control the level of quality being produced in the end product. ACI 121R-08 defines QC as actions taken by an organization to provide control and documentation over what is being done and what is being provided so that the applicable standard of good practice and the contract documents for the work are followed. According to Juran (2010), QC's main purpose is maintaining control. Performance is evaluated during operations, and performance is compared to targets during operations. In the process, metrics are utilized to monitor adherence to standards.

Random sample is a sample obtained such that all portions of the population have an equal chance or probability of being selected. For statistical evaluations, it is important that the sampling be random to ensure that no particular portion is selected more often than others, which can cause the results to be biased. In some cases, procedures for random sampling may be modified to ensure that the sample is representative; concrete from a truck mixer should be sampled from the middle portion of the load.

Range is the difference between the highest and lowest values in a series of results.

Required average compressive strength is the targeted average compressive strength of concrete used in mixture proportioning to ensure a high likelihood that the concrete will meet specified strength acceptance criteria. It is denoted as f'_{cr}.

Sample is a subset of data from the population. *Examples*: A portion of concrete sampled from a single ready mixed concrete truck for testing, Strength of three cores from a column. The test results from a sample are used to *estimate* the characteristics of the population. The thickness of the entire pavement is estimated from the average thickness of 100 cores (samples) obtained from the pavement.

Specified compressive strength is the compressive strength of concrete used in design. It is denoted as f'_c.

Standard deviation is a calculated value that provides an indication of the variation from the mean value of the data set. It is denoted as S. The formula for the calculation is given in the chapter on Basic Statistics.

Stratified random sampling is used to make sure that samples are obtained from different sections of a population. In this case, the population is divided into sublots and the random samples are obtained from within each sublot. A 100 ft length of pavement may be divided into 10 sublots, and one core will be obtained randomly from within each 10 ft length of pavement.

Strength test result is the average compressive strength of two or more cylinders made from the same sample of concrete and tested at the same age. A single cylinder strength result does not constitute a test result. According to ACI 318, at least three 4 × 8 in. (100 × 200 mm) or two 6 × 12 in. (150 × 300 mm) cylinders must be averaged for a single test result. Some specifying agencies, however, require only two 4 × 8 in. (100 × 200 mm) cylinders.

Within-batch coefficient of variation is calculated based on the average range and average strength of 10 consecutive strength test results. It is denoted as V_1. Within-batch range can be used to estimate testing variability and is discussed in the chapter on Variation Due to Testing.

Within-batch range is the difference between the maximum and minimum strengths of individual concrete specimens that comprise one strength test result. The within-batch range is typically expressed as a percent of the strength test result for that batch. Within-batch range can be used to estimate testing variability and is discussed in Chapter 10, Variation in Concrete Performance Due to Testing.

Appendix A—Submittal Requirements and Acceptance According to the ACI 318 Building Code

ACI 318—*Building Code Requirements for Structural Concrete* and ACI 301—*Standard Specification for Structural Concrete* contains detailed requirements for judging the adequacy (or acceptability) of test results. They also contain specific requirements and procedures for the mixture proportions proposed for use in the project. These mixture submittal requirements and approval procedures are considered necessary to ensure that the concrete furnished will actually meet the strength requirements. ACI 318, the building code, provides a basis for arriving at the mixture proportions to obtain required strength and durability in a structure and a process for checking the quality of concrete from test results during and after placement. The primary intent of the building code is to protect the safety of the public. ACI 301 is a model specification for concrete construction and is frequently invoked in job contract documents either in its entirety or in part. The basic provisions of ACI 301 and ACI 318 are essentially similar, but there are some minor differences in details, and the concrete producer should make sure to read the job specification to know what is required.

Note that for strength acceptance, a "test result" is the average of standard-cured cylinders—at least two 6 × 12 in. or three 4 × 8 in. cylinders breaks. The cylinders should have been prepared from the same sample of concrete obtained in accordance with procedures for obtaining samples of concrete for tests outlined in ASTM C 172.

ACI uses a statistical basis to establish the target average strength of a concrete mixture based on the specified strength the design engineer uses to design the structure. The required average strength, f'_{cr}, that the concrete mixture needs to attain is always higher than the specified strength, f'_c. This ensures that the strength tests have a low probability of falling below the specified strength. The acceptance criteria, discussed later, are based on these mixture approval procedures. The steps in the procedure can be outlined as follows:

1. If a similar mixture has been used in previous jobs, determine the expected standard deviation (S) from past test records.
   a. This is done by submitting a past record of at least 30 consecutive tests made on a similar mixture with similar materials and conditions of production. The specified strength of the concrete mixture represented by the test records should be within 1000 psi (7 MPa) of specified strength (f'_c) for the proposed work.

TABLE 5.3.1.2
Modification Factor for Standard Deviation When Less Than 30 Tests Are Available (from ACI 318)

No. of Tests[a]	Modification Factor for Standard Deviation[b]
Less than 15	Use Table 5.3.2.2
15	1.16
20	1.08
25	1.03
30 or more	1.00

[a] Interpolate for intermediate numbers of tests

[b] Modified standard deviation to be used to determine required average strength, f'_{cr}

b. If it is difficult to find a past single job with 30 tests, the standard deviation can be computed for two jobs totaling 30 or more tests. In this case, the standard deviations are computed separately for each job and then statistically averaged. This option can only be used if the total number of tests from the two records is 30 or more.

$$\text{Statistical average of two standard deviations, } \bar{S} = \sqrt{\frac{(n_1 - 1)(S_1)^2 + (n_2 - 1)(S_2)^2}{(n_1 + n_2 - 2)}}$$

where n_1 and S_1 are the number of tests and standard deviation from Job 1, respectively, and n_2 and S_2 are the number of tests and standard deviation from Job 2.

c. If a past record of 15–29 tests (from 1 job) on a similar mixture is available, then the standard deviation should be increased by the modification factor from Table 5.3.1.2. In this case, the set of test data should represent a single record of consecutive tests that span a period of not less than 45 calendar days.

2. Determine the required average strength for the concrete mixture.
   a. If a past test record with at least 15 tests is available, use the standard deviation (S) from step 1(a), 1(b), or 1(c).

      The required average strength, f'_{cr}, will be the larger of either Equation 5.1 or 5.2.

$$f'_{cr} = f'_c + 1.34\ S \tag{5.1}$$

$$f'_{cr} = f'_c + 2.33\ S - 500 \tag{5.2}$$

(In SI units for MPa, Equation 5.2 is equivalent to $f'_{cr} = f'_c + 2.33\ S - 3.5$.)

For specified strength greater than 5000 psi (35 MPa), use the larger of either Equation 5.1 or 5.2.

$$f'_{cr} = 0.90f'_c + 2.33\ S \quad (5.3)$$

Of Equations 5.1 and 5.2, Equation 5.1 will govern when the standard deviation, S, is less than 505 psi [3.5 MPa].

If the variability is known by the coefficient of variation (V) instead of S, the above three equations can be rewritten in terms of V as follows:

$$f'_{cr} = \frac{f'_c}{(1 - 1.34 \times V)} \quad (5.1a)$$

$$f'_{cr} = \frac{0.9 \times f'_c}{(1 - 2.33 \times V)} \quad (5.1b)$$

$$f'_{cr} = \frac{(f'_c - 500)}{(1 - 2.33 \times V)} \quad (5.1c)$$

b. If no test records (or less than 15 tests) are available to calculate the standard deviation, the required average strength (f'_{cr}) shall be determined from Table 5.3.2.2. This option generally will require the concrete producer to design a mixture for a higher strength.

3. The concrete producer as part of the submittal for the proposed job should furnish data to demonstrate that the proposed concrete mixture proportions will produce the required average strength needed. This can be done by either one of the following methods.
   a. A record of between 10 and 30 consecutive tests of field concrete. This would generally be the same test record that was used to document the standard deviation, but it could also be a different set of test results. It

TABLE 5.3.2.2
Required Average Compressive Strength When Data Are Not Available to Establish a Standard Deviation (from ACI 318)

US Customary Units		SI (Metric) Units	
Specified Compressive Strength, f'_{cr}, psi	**Required Average Compressive Strength, f'_{cr}, psi**	**Specified Compressive Strength, f'_c, MPa**	**Required Average Compressive Strength, f'_{cr}, MPa**
Less than 3000 psi	$f'_c + 1000$	Less than 21	$f'_c + 7.0$
3000 to 5000	$f'_c + 1200$	21 to 35	$f'_c + 8.3$
Over 5000	$1.10\ f'_c + 700$	Over 35	$1.10\ f'_c + 5.0$

could also be a combination of tests from more than one job made with the same class of concrete.

b. A series of laboratory trial batches in accordance with ASTM C192. A three-point curve with w/cm ratio or cementitious content that encompasses the required average strength, f'_{cr}, should be developed. The trial mixtures should be designed to produce a slump within ±0.75 in. of maximum permitted and an air content for air-entrained concrete within ±0.5 percent of the maximum permitted.

4. When past test records for documenting average strength potential do not exist or the size of the project does not justify trial batch evaluation, the code provides for an alternative to allow the job to proceed. "Other experience or information" may be used to achieve the concrete mixture proportions and the required average strength should be at least 1200 psi greater than the specified strength. This provision requires the engineer's permission and is not permitted for specified strength exceeding 4000 psi.
5. Data collected during the course of the job can be used to reevaluate the required average strength, f'_{cr}, when more than 15 test results have been accumulated. This can be useful if the required average strength was determined from Table 5.3.2.2, due to lack of strength records, or if the standard deviation of the current job has significantly changed.

EVALUATION AND ACCEPTANCE OF CONCRETE ACCORDING TO ACI 318

The code requires that certified technicians perform all tests on fresh concrete at the job site and on hardened concrete in the laboratory. It also indicates that laboratories conducting acceptance testing should be accredited or inspected for conformance with the requirements of ASTM C1077. It also suggests that test reports should be promptly distributed to the various parties to allow for timely determination of compliance or for corrective action.

Acceptance of concrete is based on samples of concrete obtained in accordance with ASTM C172, test specimens prepared and cured in accordance with ASTM C31, and tested in accordance with ASTM C39. The test age for cylinder strengths is 28 days or as required by the job specification. Once the mixture is approved and the job has started, the tests of standard-cured specimens made from the job concrete must satisfy *both* of the following requirements before the concrete is considered acceptable:

1. The average of all sets of three consecutive strength tests equal or exceed the specified strength (f'_c).
2. No single strength test (average of two cylinders) falls below the specified strength (f'_c) by more than 500 psi [3.5 MPa] if f'_c is 5000 psi (35 MPa) or lower, or falls below f'_c by more than 10% if f'_c is greater than 5000 psi (35 MPa).

These acceptance criteria are related to Equations 5.1–5.3 used to calculate the required average strength. If the concrete mixture is meeting the target required average strength, with a similar standard deviation of test results as that of the test record used in the submittal, the statistical probability of failing to meet both the acceptance criteria is 1 in 100. The building code indicates that, if concrete fails to meet these criteria, steps should be taken to increase the average of the subsequent test results. This might mean that a new target strength level is necessary. The building code suggests that the average level of strength can be increased by

- Increasing the cementitious material content,
- Changing the mixture proportions,
- Better control on the slump and air content,
- Reduction in the delivery time, and
- An improvement in the testing, including evaluating whether the standard procedures are being followed.

When a strength test fails the second criterion, the building official will be concerned about the safety of the structure and a more thorough evaluation of the portion of the structure represented by the low strength result is necessary. This might include reevaluation of the f'_c needed in that portion, core testing, or load testing.

EXAMPLE OF ACI REQUIREMENTS

The specified strength (f'_c) for a job is 3000 psi or 21 MPa. The standard deviation (S) calculated from 30 tests of the same class of concrete produced from the same plant is 650 psi or 4.5 MPa.

The required average strength (f'_{cr}) is determined from the larger value calculated from the ACI equations:

$$f'_{cr} = f'_c + 1.34\ S$$

$$f'_{cr} = 3000 + (1.34 \times 650) = 3871\ \text{psi}$$

$$f'_{cr} = 21 + (1.34 \times 4.5) = 27.03\ \text{MPa}$$

$$f'_{cr} = f'_c + 2.33\ S - 500$$

$$f'_{cr} = 3000 + (2.33 \times 650) - 500 = 4015\ \text{psi}$$

$$f'_{cr} = 21 + (2.33 \times 4.5) - 3.5 = 27.99\ \text{MPa}$$

The concrete mixture should be designed to produce an average strength of at least 4015 psi or 27.99 MPa.

Appendix B—Calculation of Strength Variation from Water Content Variation

The ACI 211.1 table of strength versus w/cm for non air-entrained concrete has been plotted below with the best-fit straight lines for both the inch-lb and metric units.

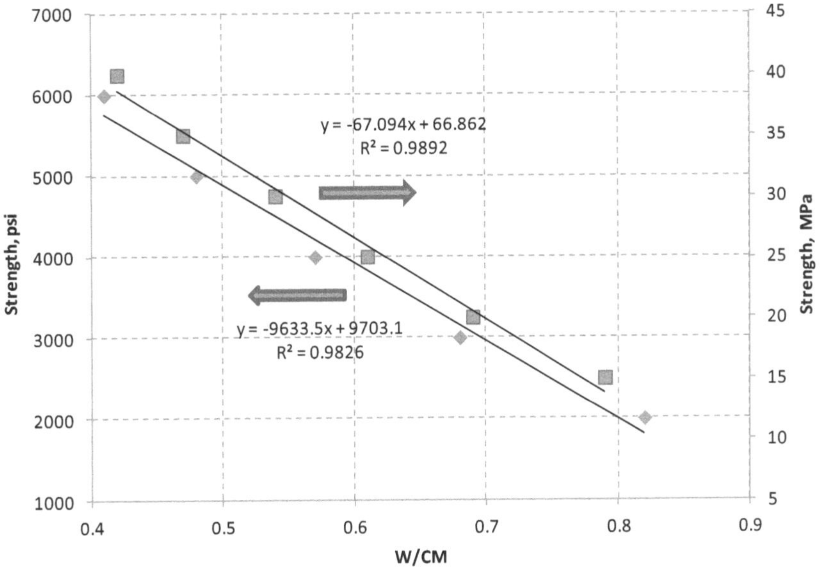

The slope is -9633 psi and -67 MPa. The following table shows the calculated change in strength (based on the straight line equation) for a one pound (kilogram) change in cementitious content for concrete with a mixing water content of 300 lb/yd^3 (180 kg/m^3) and for different target w/cm values. The numbers would be slightly higher for lower water content.

w/cm	Change in Strength, psi/lb of Cement	Change in Strength, MPa/kg of Cementitious
0.4	5	0.06
0.5	8	0.09
0.6	12	0.14
0.7	16	0.18

It is seen that the strength change for a change in one pound of cementitious is 8-12 psi (0.09-0.14 MPa) for concrete in the 4000 to 5000 psi (27 to 33 MPa) range and for w/cm of 0.50 to 0.60.

Using the straight line equation, the percentage change in strength for a ±5% change in cementitious or water content for concretes for a given water content is calculated and is shown below:

w/cm	Change in Strength, %
0.4	3.3%
0.5	4.9%
0.6	7.4%
0.7	11.4%

Thus it is clear that a 5% change in cementitious content will result in a 5% change in compressive strength only around w/cm of 0.50.

Higher w/cm concrete mixtures appear to be more vulnerable to cementitious and water content variations as compared to lower w/cm concrete mixtures. A given percent change in cementitious or water contents does not lead to the same percent change in compressive strength because strength is related inversely to water-cementitious ratio, and not directly to cementitious or water content as commonly assumed.

Appendix C—FLYASH Tests (NRMCA Draft Standard Practice 1988)

UNIFORMITY OF COLOR

An approximately 1 in. (25 mm) thick layer of fly ash shall be carefully placed on top of the previous sample of ash in a 1000 mL hydrometer jar (ASTM D422). Rate the color of the current sample in comparison to the immediately preceding sample according to the following scale.

Grade 2	Description
−2	Much lighter, a color difference clearly detected
−1	Slightly lighter, color difference detectable by most observers
0	Same, no color difference by at least half of a group of four observers
+1	Slightly darker, color difference detectable by most observers
+2	Much darker, a color difference clearly detected

FOAM INDEX TEST (MEININGER 1981)

The foam index test measures rapidly the effect of a fly ash sample on the required air-entraining admixture dosage to obtain the required entrained air content in a concrete mixture and will help detect a change in fly ash properties from previous shipments. The foam index test has not been standardized by ASTM. The procedure described here has been used at NRMCA since later 1970s. Place 16 grams cement +4 grams fly ash in a wide-mouth glass bottle. Add 50 mL water, cap bottle, and shake for 1 minute. Add air-entraining agent (diluted 1:20 with water) in measured increments using an accurate pipette. After each addition, cap and shake vigorously for 15 seconds. Remove cap and observe the stability of the foam. The amount of diluted air-entraining agent needed to produce a stable foam that just covers the surface is the foam index of the fly ash. The foam index test can also be run on 20 grams of cement alone to understand the influence of cement shipment. The foam index test might also be run, including 40 grams of sand to understand the influence of sand shipments.

MORTAR AIR CONTENT AND AIR LOSS (LANE 1991) (MODIFIED C311)

ASTM C311 suggests using air entraining admixture (AEA) dosage to target mortar air content of 18%. Research had indicated (Meinenger [1981]) that to attain 18%

mortar air content, an excessive amount of AEA was required, which was ineffective in evaluating the effect fly ash had on air entrainment. Therefore, an AEA dosage to target mortar air content between 10% and 14% was used for all fly ash samples from that source. The AEA is added to the mixing bowl while the sand is being introduced, and the rest of the experimental steps are similar to C311. After weighing the 400 mL measure containing the mortar, the mortar air content was calculated gravimetrically using the equation provided in C311. This was the initial air content. In addition, the stability of the entrained air system was measured according to the procedure below:

Immediately after weighing the 400 mL measure the mortar, including the mortar used for flow determination, was returned to the mixing bowl. The bowl was covered to prevent evaporation. After a 45-minute rest period (from the time of mixing), the bowl was uncovered and remixed at medium speed for 5 minutes. The target flow of 80% to 95% was attained with some water adjustments after which the mortar air content was measured again gravimetrically by weighing the mortar-filled 400 mL measure. This was the final mortar air content. The difference between the initial and final air content provided information about the stability of the entrained air content.

Appendix D—Initial Curing of Acceptance Test Cylinders and Transportation (NRMCA ASCC Checklist)

1. Initial curing (up to 48 h)
   - Immersed in water-controlled temperature
   - Storage box-controlled temperature; record daily minimum and maximum temperature

Note: In the absence of cylinder storage with daily record of minimum/maximum temperatures, cylinders shall be immersed in water immediately after molding

2. Responsibility for providing cylinder storage box
3. Responsibility for maintaining temperature in storage box between 60° and 80°F (16°C–27°C) during first 16–48 hours after molding
4. Describe how storage box temperature will be maintained
5. When will cylinders made on days preceding nonwork days be transported to the laboratory?
6. Describe arrangements for access to construction site on nonwork days
7. Responsibility for final curing as per ASTM C31

Appendix E—Outline of Training Requirements

QUALITY CONTROL/QUALITY ASSURANCE MANAGERS

In lieu of certifications, the following outlines the minimum training for QA/QC manager. These personnel should have had at least 36 hours of training:

Fundamentals of quality concrete	Batching accuracy requirements and controls
Types of concrete for different applications and concrete materials required	Requirements for concrete in ACI 318, ACI 301 and state specifications
Ingredient materials, applicable specifications, and material certifications	Basic statistical concepts for quality control measurements and monitoring
Aggregate tests and associated calculations	General quality control procedures and testing
Mixture proportioning and adjustments to mixtures for yield, moisture, changes in material characteristics	Hot and cold weather concrete and adjustments to mixtures
Fresh concrete properties and test methods	Handling, placing, and finishing concrete and applicable requirements for concrete, depending on methods used
Hardened concrete properties and test methods	Management of returned concrete and impact on quality
Fundamentals of durability and requirements for concrete	Troubleshooting concrete problems and defects
Concrete production and plant operations—requirements of ASTM C94	Ability to establish and train personnel on adjustments to concrete mixtures at the job site

The QA/QC personnel should have documented at least 8 hours of annual continuing education on this subject matter. Attendance of national and local technical training, conventions, and committee meetings are acceptable.

PLANT OPERATOR

In lieu of certifications, the following outlines the minimum training for plant operators. Plant operators should have had at least 8 hours of training:

Basic concrete technology	Intra-plant material handling and storage
Types of concrete and concrete materials	Batching sequence and control
Aggregate moisture tests and adjustments	Daily plant startup and shutdown checklist
Effects of changes in materials to mixtures	Plant maintenance

continued

(Continued)

Control of slump and air	Inventory management of materials
Basics of mixture proportioning	Identification of out-of-spec loads and disposition
Fresh concrete tests and properties	Yield adjustment
Production—general concepts and plant operation	Temperature control of concrete
Accuracy requirements for scales, meters	Order handling and delivery tickets
Batching accuracy requirements and controls	Plant safety

The plant operator should have documented at least 4 hours of annual training on this subject matter.

TRUCK MIXER OPERATOR

In lieu of certifications, the following outlines the minimum training for truck mixer operators. Truck mixer operators should have had at least 8 hours of training:

Basic concrete technology	Procedures for placing and finishing concrete
Types of concrete and concrete materials	Factors impacting concrete due to ambient temperature
Mixing requirements, water additions	Handling and recording customer complaints
Control of slump and air	Delivery tickets and job-site notes
Company policy on job-site water additions	Environmental regulations—delivery and job site
Truck and mixer operation, maintenance and production of concrete	Safety—driving and personal
Fresh concrete test methods and procedures—recognition of improper procedures	Handling job-site concrete rejection

Truck mixer operator should have documented at least 4 hours of annual training on this subject matter.

Appendix F—Checklist for Laboratories Used for Mixture Development

1. The laboratory shall maintain documentation of the following:
   1.1 Laboratory equipment inventory.
   1.2 Personnel qualifications.
2. The laboratory is under the direction of a licensed professional engineer with at least 3 years experience in materials testing or an individual with at least 7 years experience in concrete technology and materials testing.
3. Concrete laboratory technicians possess current technician certification that covers ACI Field Grade I and ACI Strength Testing Technician certification. ACI Lab Level 1 is an acceptable alternative.
4. Laboratory maintains current published annual book of ASTM or other standards on tests that it performs.
5. Internal quality audits for the lab are conducted annually, and records are maintained.
6. Curing tank/room requirements:
   6.1 Curing tank/room that can be used for curing test specimens and meets the requirements stated in ASTM C511.
7. Documentation of equipment
   7.1 Verification maintained in the laboratory.
   7.1 Verification documented annually for scales, balances, slump cone (ASTM C143), air content (C173), temperature (C1064), dimensions of cylinder molds (C470); sieves checked for defects. Frequency at every 3 months for pressure air meters (C231), sulfur capping (C617).
   7.2 Compressive strength machine should conform to C39. Its calibration is annually verified.

Appendix G—Preventive Equipment Maintenance

SILOS FOR CEMENTITIOUS MATERIALS

- Check tightness of separation walls in multicompartment silos by determining accumulation of cementitious materials in compartment left in "empty" condition.
- Check for accidental "cross-feed" in transfer devices.
- Monitor the high-bin indicators, anti-overfill devices and pressure sensors for correct operations.
- Inspect the dust collection systems in accordance with the manufacturer's recommendations.

AGGREGATES

- Periodically empty and inspect the wear of the overhead aggregate bins.
- Monitor transfer devices, turnhead limit switches, and full-bin signals for correct operation.
- Process to verify product type in different storage bins.

CHEMICAL ADMIXTURES

- Inspect agitation devices as required to maintain uniform solution densities (e.g., standard calcium chloride solutions). Tanks, hoses, and dispensers for seasonal admixtures should be operated weekly.
- Ensure that admixture dispensers are functioning correctly. Make sure that sight glasses are clean and are provided with legible graduations. Periodically check for tank integrity and for leaking hoses and faulty connections.

BATCHING EQUIPMENT

- Make sure that weigh batchers remain freely suspended; that the scale linkages are clean; that wind protection is adequate for cement weigh batchers; there is no binding against the frame or other obstructions (check when weighing capacity loads); and there is no binding of scale cables (at entry port to control house).
- Ensure that the cement weigh batcher is properly vented, and there is no back pressure on the scale system from pneumatically charging the cement or pozzolan silos (aerator; or from pneumatic unloading of transportation units).

- Rotate the sheaves or cable pulleys periodically for uniform wear.
- Discuss record maintenance policies relative to time and cross reference to delivery tickets.
- Make provision for periodically checking the actual admixture discharge into mixer. Obstructions at end of discharge line, or low air pressure in pneumatic discharge, may cause holdback that may enter next batch.

CENTRAL MIXER

- Designate the frequency of checking for buildup of hardened concrete and blade wear.
- Define the frequency of performing mixer uniformity test, if conducted.
- Check mixing time to programmed timer.

TRUCK MIXERS

- Check for concrete buildup, blade wear, functioning revolution counter, accuracy of water measuring device, and general condition of truck water system (legible quantity indications; clean gauge, water leaks).
- Verify the TMMB rating plate for mixer and agitator capacity and manufacturer's plate for operating details.
- On units with hydraulic slump meters, verify correlation of hydraulic pressure to slump of standard mixes and load size established.

Index

A

B

C

D

L

M

N

O

P

W

X

Y

Z